Preparing for a New Calculus

Conference Proceedings

Preparing for a New Calculus

Conference Proceedings

Anita E. Solow, Editor

MAA Notes Number 36

Published and Distributed by
The Mathematical Association of America

The conference, "Preparing for a New Calculus," was supported by the National Science Foundation through the Division of Undergraduate Education; the Division of Elementary, Secondary and Informal Education; and the Division of Mathematical Sciences.

ISBN 0-88385-092-3
Library of Congress Catalog Number 94-77348
Printed in the United States of America
Current Printing
10 9 8 7 6 5 4 3 2 1

MAA Notes and Reports Series

The MAA Notes and Reports Series, started in 1982, addresses a broad range of topics and themes of interest to all who are involved with undergraduate mathematics. The volumes in this series are readable, informative, and useful, and help the mathematical community keep up with developments of importance to mathematics.

MAA Notes

1. Problem Solving in the Mathematics Curriculum, *Committee on the Teaching of Undergraduate Mathematics,* a subcommittee of the Committee on the Undergraduate Program in Mathematics, *Alan H. Schoenfeld,* Editor
2. Recommendations on the Mathematical Preparation of Teachers, *Committee on the Undergraduate Program in Mathematics, Panel on Teacher Training.*
3. Undergraduate Mathematics Education in the People's Republic of China, *Lynn A. Steen,* Editor.
5. American Perspectives on the Fifth International Congress on Mathematical Education, *Warren Page,* Editor.
6. Toward a Lean and Lively Calculus, *Ronald G. Douglas,* Editor.
8. Calculus for a New Century, *Lynn A. Steen,* Editor.
9. Computers and Mathematics: The Use of Computers in Undergraduate Instruction, *Committee on Computers in Mathematics Education, D. A. Smith, G. J. Porter, L. C. Leinbach, and R. H. Wenger,* Editors.
10. Guidelines for the Continuing Mathematical Education of Teachers, *Committee on the Mathematical Education of Teachers.*
11. Keys to Improved Instruction by Teaching Assistants and Part-Time Instructors, *Committee on Teaching Assistants and Part-Time Instructors, Bettye Anne Case,* Editor.
13. Reshaping College Mathematics, *Committee on the Undergraduate Program in Mathematics, Lynn A. Steen,* Editor.
14. Mathematical Writing, by *Donald E. Knuth, Tracy Larrabee, and Paul M. Roberts.*
15. Discrete Mathematics in the First Two Years, *Anthony Ralston,* Editor.
16. Using Writing to Teach Mathematics, *Andrew Sterrett,* Editor.
17. Priming the Calculus Pump: Innovations and Resources, *Committee on Calculus Reform and the First Two Years,* a subcomittee of the Committee on the Undergraduate Program in Mathematics, *Thomas W. Tucker,* Editor.
18. Models for Undergraduate Research in Mathematics, *Lester Senechal,* Editor.
19. Visualization in Teaching and Learning Mathematics, *Committee on Computers in Mathematics Education, Steve Cunningham and Walter S. Zimmermann,* Editors.
20. The Laboratory Approach to Teaching Calculus, *L. Carl Leinbach et al.,* Editors.
21. Perspectives on Contemporary Statistics, *David C. Hoaglin and David S. Moore,* Editors.
22. Heeding the Call for Change: Suggestions for Curricular Action, *Lynn A. Steen,* Editor.
23. Statistical Abstract of Undergraduate Programs in the Mathematical Sciences and Computer Science in the United States: 1990–91 CBMS Survey, *Donald J. Albers, Don O. Loftsgaarden, Donald C. Rung, and Ann E. Watkins.*
24. Symbolic Computation in Undergraduate Mathematics Education, *Zaven A. Karian,* Editor.
25. The Concept of Function: Aspects of Epistemology and Pedagogy, *Guershon Harel and Ed Dubinsky,* Editors.
26. Statistics for the Twenty-First Century, *Florence and Sheldon Gordon,* Editors.
27. Resources for Calculus Collection, Volume 1: Learning by Discovery: A Lab Manual for Calculus, *Anita E. Solow,* Editor.
28. Resources for Calculus Collection, Volume 2: Calculus Problems for a New Century, *Robert Fraga,* Editor.
29. Resources for Calculus Collection, Volume 3: Applications of Calculus, *Philip Straffin,* Editor.
30. Resources for Calculus Collection, Volume 4: Problems for Student Investigation, *Michael B. Jackson and John R. Ramsay,* Editors.

31. Resources for Calculus Collection, Volume 5: Readings for Calculus, *Underwood Dudley,* Editor.
32. Essays in Humanistic Mathematics, *Alvin White,* Editor.
33. Research Issues in Undergraduate Mathematics Learning: Preliminary Analyses and Results, *James J. Kaput and Ed Dubinsky,* Editors.
34. In Eves' Circles, *Joby Milo Anthony,* Editor.
35. You're the Professor, What Next? Ideas and Resources for Preparing College Teachers, *The Committee on Preparation for College Teaching, Bettye Anne Case,* Editor.
36. Preparing for a New Calculus: Conference Proceedings, *Anita E. Solow,* Editor.

MAA Reports

1. A Curriculum in Flux: Mathematics at Two-Year Colleges, *Subcommittee on Mathematics Curriculum at Two-Year Colleges,* a joint committee of the MAA and the American Mathematical Association of Two-Year Colleges, *Ronald M. Davis,* Editor.
2. A Source Book for College Mathematics Teaching, *Committee on the Teaching of Undergraduate Mathematics, Alan H. Schoenfeld,* Editor.
3. A Call for Change: Recommendations for the Mathematical Preparation of Teachers of Mathematics, *Committee on the Mathematical Education of Teachers, James R. C. Leitzel,* Editor.
4. Library Recommendations for Undergraduate Mathematics, *CUPM ad hoc Subcommittee, Lynn A. Steen,* Editor.
5. Two-Year College Mathematics Library Recommendations, *CUPM ad hoc Subcommittee, Lynn A. Steen,* Editor.

These volumes may be ordered from:
The Mathematical Association of America
1529 Eighteenth Street, NW
Washington, DC 20036
800-331-1MAA FAX 202-265-2384

Introduction
The Fruits of Success: Calculus and Precalculus Reform in 1993

Anthony Peressini
University of Illinois

Origins

Many promising projects exploring new content and teaching methods for mathematics at the high school and beginning college levels are currently under way or have been completed. Many of these are associated with various national agendas for change, including the calculus reform movement, the implementation of the NCTM *Standards,* and the discrete mathematics and educational technology initiatives.

The national movement to reform calculus was launched at a conference at Tulane University in 1986. The movement blossomed quickly, nurtured by the publication of the conference proceedings, *Toward a Lean and Lively Calculus* [1], the subsequent national report *Calculus for a New Century* [3], and numerous journal articles. (See David Smith's article "Trends in Calculus Reform," elsewhere in this volume, for an extensive bibliography.)

In 1988, the National Science Foundation initiated its Curriculum Development in Mathematics: Calculus program to support course development in calculus. This program funded many projects, some of which are sure to obtain fundamental changes in the future content and mode of instruction of the calculus courses in our high schools and colleges. Although the primary setting for these developmental projects has been colleges, several are being tested in high schools as well.

A reshaping of precalculus mathematics at the college level (i.e., community colleges, colleges and universities) has been spawned both by the critical reexamination of calculus now under way as well as by the fact that a disappointingly small percentage of students in the traditional college precalculus course complete calculus. Precalculus at the college level has been a remedial course, either a review of mathematics taken earlier but not retained or an intensive presentation of mathematics that should have been learned earlier.

A parallel reform movement in mathematics instruction at the precollege level is also under way. Like the calculus reform movement, it had its beginnings in national reports that identified serious problems with the current state of affairs. The NCTM *Curriculum and Evaluation Standards for School Mathematics* [2], published in 1989, provided specific, realistic national goals for this reform, but the implementation of the *Standards* was left to state and local initiatives and to developmental projects in curriculum revision, teacher training and instructional methods. Although the primary setting for those reform efforts driven by the NCTM *Standards* has been at the K–12 level, colleges are playing a major role in the teacher preparation and precalculus components of this reform.

Precalculus at the high school level must either be interpreted much more broadly than it is at the college level, or it must be regarded as one component of the college preparatory curriculum that may include other major components such as discrete mathematics, statistics and probability. Precalculus at the high school level is not restricted to a single course or pair of courses as it is in college. Precalculus is not a remedial course in high school; it is one capstone for a good high school mathematics program.

The reform movements in precalculus and calculus at the high school and college levels are influencing, and being influenced by, several current developments including:

- the strong interest in and support for the Advanced Placement (AP) Calculus Program that exists among parents, teachers, and students in our nation's high schools;
- the strong support for earlier and more extensive training in discrete mathematics, statistics, and mathematical modeling that exists among mathematics teachers and mathematics "user" groups;
- the movement toward greater use of available technology to improve mathematics instruction at all levels;

- college admissions and mathematics placement testing instruments and policies that tend to inhibit curricular reform.

These influences, as well as the deficiencies of the traditional calculus and precalculus courses, have prompted mathematicians and mathematics educators to craft a variety of strikingly new courses using new teaching strategies. Most of these courses are under development, and their implementation (beyond pilot testing sites) is just beginning. Nevertheless, it is possible to recognize some common themes in these courses and to begin to analyze the interconnections between the new calculus courses and their precalculus counterparts.

What Is Different about New Calculus and Precalculus Courses?

For one thing, it is unlikely that a single new course in calculus or precalculus will emerge triumphant as a replacement for the traditional monoliths. Rather, several of these courses will survive as viable and distinct alternatives, and more than one of these courses will be offered at some institutions.

These predictions rest on the current critical reexamination of both calculus and precalculus that leads to development of several successful course models quite different in their content and instructional approaches. Although there is broad agreement concerning general content and instructional strategies for these new courses, there are substantial differences in specific implementations and a healthy realization that more than one approach is not only possible but even desirable.

As noted earlier, the new calculus courses have been developed primarily at the college level, while the new precalculus courses have been designed independently and for differing reasons at the high school and college levels. Given these differences in origin, it is surprising and heartening that there is broad agreement among the developers on the characteristics distinguishing these courses from their traditional counterparts. These characteristics are as follows:

- The new courses place greater emphasis on the meaning of mathematical concepts and the use of mathematical methods to analyze problems. They de-emphasize manipulative skills and the application of formulas or rules. They cover fewer subsidiary topics.

- The new courses place greater responsibility for learning in the hands of the students by requiring them to read more critically, to work independently or in small groups with less prescriptive guidance from the teacher and text, to analyze problems, and to explain their solutions. They also provide fewer "template" solutions for use by the students in doing homework, and they place less emphasis on the classification of problems by means of some preferred solution method.

- The new courses typically use a multiple approach to the analysis of new concepts and problems; investigating them systematically from the numerical, graphical, and algebraic perspectives. In traditional courses, the algebraic approach predominated, and graphical considerations often became an end in themselves rather than a means for understanding problems and concepts.

- The new courses place greater emphasis on mathematical modeling and the solution of realistic problems as a means for motivating mathematical concepts and methods and for applying them.

- The graphical, numerical, and symbolic capabilities of computers and calculators are used systematically as tools that sometimes drive the presentation of the content in the new courses. In the traditional courses, available technology is either ignored or is an optional "add-on" to the presentation.

Of course, these characteristics appear in varying degrees in the new courses, but the measure of their cumulative presence is perhaps an indication of the extent to which a new course is truly a "reform" course.

The Objectives, Program, and Outcomes of The Conference/Workshop "Preparing for a New Calculus"

I was assisted by Don Albers in organizing the conference/workshop "Preparing for a New Calculus," which was held at the University of Illinois Allerton Conference Center on April 22–25, 1993. The National Science Foundation provided travel and subsistence support for the participants as well as the funds for the publication and limited distribution of the conference proceedings. Its Planning Committee advised on the selection of participants, the program, and other matters. That committee consisted of:

Frank Demana, Ohio State University
John Dossey, Illinois State University

Wade Ellis, West Valley Community College (California)
John Goebel, North Carolina School of Mathematics and Science
Sue Ann Mc Graw, Lake Oswego High School (Oregon)
Sheldon Gordon, Suffolk County Community College (New York)
Deborah Hughes Hallett, Harvard University
David Smith, Duke University
J. Jerry Uhl, University of Illinois
Zalman Usiskin, University of Chicago

The conference/workshop sought to identify the emerging themes of the new calculus and to explore the implications of these themes for precalculus education in high school and college as well as those for subsequent courses at the college level. A further purpose was to explore curricular and methodological changes in teaching high school and college precalculus mathematics and the potential these changes hold for entry level college courses including calculus. Among the questions considered were the following:

- What are the content and methodological themes of the new calculus courses?
- What specific changes in precalculus course content and methodology would help students succeed in these new calculus courses?
- How do recent curricular changes at the high school level relate to the needs of students entering the new calculus courses?
- How does the AP Calculus Program fit in with the new calculus courses?
- How do the new calculus courses relate to the curricular objectives of providing earlier training in discrete mathematics and statistics?
- What are the implications of the new calculus courses for the training of high school mathematics teachers and university teaching assistants and in the preparation of graduate students in mathematics for college teaching?
- What are the implications for college mathematics placement testing of the new calculus and precalculus courses?

Eighty individuals were invited to the conference. The participants were selected from the following groups:

a. Those working on calculus reform developmental and implementation projects;

b. Those engaged in precalculus initiatives at the college level and college preparatory projects at the high school level;

c. Individuals directly involved in initiatives tied to calculus and precalculus such as the Advanced Placement Calculus Programs, the discrete mathematics and instructional technology initiatives, and teacher training at the high school and college level;

d. Representatives of national mathematical organizations such as MAA, NCTM, AMS, AMATYC, NSF, and MSEB; and

e. Representatives of textbook publishers and educational technology providers.

Many of the individuals invited to the conference represented two or more of these groups. Forty of the participants were primarily associated with 4-year college and university projects, twenty-five were associated with projects directed toward the high school level, seven represented community college projects, and eight were primarily associated with groups d) and e) above. For a sense of the diverse perspective, see the complete list of participants found in the Appendix.

The seven background papers were distributed to participants prior to the conference/workshop. Among these papers were overviews of developmental projects in three areas: calculus reform (David Smith), precalculus reform at the college level (Frank Demana), and mathematics curriculum reform at the high school level (Zalman Usiskin). In addition, there were background papers on the current status and special problems and potential for curricular reform in community colleges (Sheldon Gordon and Stephen Rodi), and in colleges with large proportions of minority students (James Fife). The remaining two summarize the results of the MAA Survey of the current state of calculus reform at the college level (James Leitzel and John Dossey) and upcoming changes in the AP Calculus Program (John Kenelly and John Harvey). These papers can be found in Part 1 of this volume.

The core of the meeting program was a series of four concurrent workshops, each meeting three times with approximately twenty participants in each session. Each workshop was assigned a topic to consider in detail, but the deliberations of the workshop were not limited to that topic. The topics are listed below, together with the workshop moderators and reporters:

Topic	Moderator	Reporter
Content	Deborah Hughes Hallett	Sheldon Gordon
Teaching Strategies	John Dossey	Donald Bushaw
Institutional Context	Lee Yunker	Susanna Epp
Course Context	Carolyn Mahoney	John Mc Connell

After the second workshop session, all participants assembled in a session in which the moderator and a reporter for each workshop presented an interim summary of the deliberations of their workshop. The remainder of this general session was devoted to discussion of these summaries. The participants could join a different workshop for the third session. Following this third session, the moderators and reporters drafted a final report of their workshops for consideration at a concluding plenary session. These reports were then mailed to all participants for final comment, and finally revised to the current form found in Part 2 of these proceedings.

The workshop reports found in Part 2 give the best detailed description of the specific issues and recommendations discussed at the conference/workshop. As you will see from those reports, there was substantial agreement among the participants on the general characteristics and objectives of the new calculus and precalculus courses. The specifics of implementation are much more difficult to chart, and a number of issues were identified that remain to be resolved. On some of these issues, especially those related to course content and emphasis, there was considerable disagreement. Further development, careful evaluation, and some experience gained from the implementation of new courses will be required to resolve these issues.

Some participants contributed papers on issues related to calculus and precalculus reform. A selection of these papers can be found in Part 3. These papers have been arranged into subsections according to whether their main focus is on calculus or precalculus or calculus-precalculus combined. Although many readers are likely to be interested primarily in either calculus or precalculus, we hope that they will take this opportunity to review recent progress in the "other" field.

Part 4 of these proceedings is comprised of brief descriptions of a number of developmental projects in the calculus and precalculus reform movements. Although the list included here is by no means exhaustive, it does provide an indication of the variety of initiatives that are currently under way and is a good starting point for those who may be interested in learning more about these and other initiatives.

Apart from the specific issues and recommendations that were discussed, one of the most fruitful outcomes of the conference/workshop was the opportunity for those actively involved in high school mathematics to discuss common concerns with their college counterparts. These two groups have not had adequate opportunity for interaction in the past. It was clear that members of each group learned much about recent developments in the other setting. Further interaction between these groups should be sustained through special sessions at NCTM and MAA meetings and through occasional joint conference/workshops.

References

1. Douglas, Ronald E., ed., *Toward a Lean and Lively Calculus,* MAA Notes No. 6, 1986.

2. National Council of Teachers of Mathematics, *Curriculum and Evaluation Standards for School Mathematics,* NCTM, 1989.

3. Steen, Lynn A., ed., *Calculus for a New Century: A Pump, Not a Filter,* MAA Notes No. 8, 1987.

Contents

Part 1

Background Papers

Trends in Calculus Reform

David A. Smith
Duke University

Some two-and-a-half years ago, I was given a charge similar to the one that led to this paper: Survey the trends and common threads in calculus reform (if any), and tell us what makes a "reformed" course a *better* course. The charge came in an invitation to speak to the annual meeting of NSF Principal Investigators on calculus projects, held that year in Columbus, Ohio. The response to my remarks was sufficiently supportive that I wrote down a version of the talk (supplemented by comments from other speakers at the meeting), which was subsequently published in *UME Trends* [7]. Rereading that article, I find that I said then many of the things I want to say now. Thus, this essay is essentially an expansion and update of a two-year-old opinion piece. This is not to say that I was especially prescient—only that, at least for the last two years (probably longer), players and observers of the calculus reform game have, for the most part, agreed on the major components of "reform." The new things to say mostly fall into the categories of (a) refinements and (b) effects of more students and more faculty being brought under the reform umbrella.

I will state my positions under five general headings: (i) calculus reform as a movement; (ii) changes in content; (iii) changes in methodology; (iv) changes in students; and (v) changes in instructors. Conference participants had an opportunity to read an earlier draft of this paper prior to the conference and to comment on it in one of the opening sessions; my post-conference revision includes some of their very helpful contributions. At the end of the paper, following a brief summary-and-challenge section and a list of references, I include a bibliography on calculus reform extracted from the *Project CALC Instructor's Guide*.

The Movement

The first good thing that came out of the calculus reform initiative was the movement itself. There can be no doubt that it is indeed a movement: Witness the CRAFTY volume [9], with information on more than 70 projects that were already under way three years ago. Now, the Harvard Consortium project alone has well over 100 schools using its materials, and several other major projects report dozens of active-user sites, many with multiple sections and/or teaching all their calculus students in reformed sections. NSF has funded a study that documents the extent of reform projects and experiments already under way; a report by John Dossey and Jim Leitzel (Illinois State University and the University of Nebraska, respectively) elsewhere in this volume indicates that 60% of responding four-year schools and 45% of responding two-year schools have at least a modest reform effort in place. Rob Cole (The Evergreen State College), co-director of the Washington Center dissemination project, reports that, by actual head count, 27% of all college-level calculus students in the state of Washington are in reformed courses.

For several years now, we have seen crowded and overflowing sessions at national and regional meetings of MAA, NCTM, and AMATYC—panels, featured speakers, contributed papers, minicourses, poster sessions—on the subjects of calculus, writing, technology, and education. Sessions on these topics have been fewer in number, but just as popular, at meetings of AMS, SIAM, and AAAS. An indication of the growth in interest can be seen by comparing the programs of the annual Joint Mathematics Meetings (mostly MAA and AMS) for the years 1985 and 1993. The 1985 meeting in Anaheim was the site of the panel discussion "Calculus Instruction: Crucial but Ailing," organized by Steve Maurer (Swarthmore College) and Ron Douglas (SUNY Stony Brook), that was an important precursor of the calculus reform movement. That year, about 10 out of more than 500 events on the AMS program were related to education, plus about 28 MAA events, including 10 minicourses. (At that time, most of the invited MAA addresses were on mathematics, not on education.) This year, fully 25% of the more than 1200 numbered items on the combined AMS/MAA program were devoted to education, and the number of minicourses had grown to 17.

Participants in NSF-funded and other reform projects are receiving a steady stream of invitations to visit other campuses and help other faculties get started with reform of their own programs. Many of these projects are also running workshops, on their own campuses or at regional sites, of sufficient duration to get participants thoroughly familiar with both new materials and new methods of presentation. Growing numbers of the on-campus visitors and workshop presenters are now "second generation" reformers—not the initial developers of the materials, but users with enough experience to be ready and willing to share with newcomers. There is still a large pool of institutions whose faculties have not instituted any reforms but who are eager to begin. And most of them are finding that their administrations are also ready to commit new resources—even wondering when their mathematicians will get around to asking.

The American Mathematical Society's e-Math service now maintains a Calculus Reform bulletin board that has subscribers in the U. S., Canada, Puerto Rico, Australia, and Taiwan. Recent contributions have included lively debates on how and whether to treat series in a first-year course, whether reformers pay enough attention to mathematically talented students, whether proofs are important (relative, say, to heuristic arguments) or even possible (in the absence of axiomatic foundations), and whether the Mean Value Theorem belongs in the course. Only a few years ago, discussion of some of these matters would have been considered heretical, and resolution of some too obvious to be worth discussing.

The earliest reform projects were already under way when the NCTM *Standards* [5] appeared, and the importance of the *Standards* was not automatically noticed on college campuses. Nevertheless, almost all of the projects have converged toward a common set of goals—highly consistent with the *Standards*—and yet have maintained widely differing approaches and materials. We already have a constructive proof that there are many ways—affordable, transportable ways—to teach calculus effectively. Thus, we need never again settle into the rut of offering the same course year in and year out. An important indicator that the movement has achieved critical mass was the feeding frenzy by publishers from roughly mid-1991 to early 1992, during which time almost all the major curriculum development projects were signed to commercial publication contracts.

While there is no doubt in anyone's mind that there really is a movement for reform, some conference participants question whether it will be possible to sustain reform, and there is even talk of a "backlash" beginning to emerge. In my view, we do not yet have clear evidence for long-term sustenance, and the pressures created by reaction and inertia may well be intense. Our job is not over, but I have no doubt that reform can succeed; it's up to us to see that it does.

Course Content

Almost everyone agrees that students must constantly experience calculus concepts in a rich interplay of symbolic, numerical, and graphical forms—what the Harvard Consortium project and others have popularized as "The Rule of Three." In the last year or so, Deborah Hughes Hallett (Harvard) has been saying publicly, "We really should have called it The Rule of Four," the fourth form of representation being writing.

Every calculus reform program focuses on developing thinking skills and conceptual understanding—on eliminating the possibility of students being certified as having learned calculus when all they have demonstrated is modest proficiency at memorizing formulas and manipulating symbols. Some projects (e.g., Duke University's Project CALC) have had a strong writing component from the outset; others are now coming to the conclusion that writing is an important tool both for student learning of concepts and for faculty observation and measurement of that learning.

Several conference participants suggested a "Rule of Five," with oral communication considered on a par with the other four. To some extent, the subject of multiple representations is at least as much an issue of pedagogy as of content; we will see in the next section that these issues cannot be completely separated.

Most content developers see a need for a steady interplay between the discrete and the continuous—in the phenomena being studied, in the models of those phenomena, and in the methods for dealing with those models. Here's an example: Growth of populations and spread of epidemics are powerful motivators for understanding exponential and other forms of growth. Biological populations are inherently discrete, but sufficiently large populations with no minimal inter-reproduction time can be conveniently and accurately modeled continuously: simple, variable-separable differential equations. With sufficiently simple assumptions about the growth model (e.g., growth rate proportional to the population), solutions may be obtained by undoing differentiation formulas. In other cases (e.g., logistic growth studied before integration techniques), symbolic solutions may not be accessible, but students can generate numerical and graphical solutions via Euler's method, a discrete approximation to the model. On the one hand, Euler's method is nothing more than "rise equals slope times run," which students grasp with little difficulty. On the other hand, Euler's method is what traditional texts call "the differential approximation." When every student has a calculator in hand, it makes no sense to use differentials to approximate square roots, but it makes a lot of sense to use Euler's method to generate interesting, new, and useful functions.

A few years ago, there was a popular misconception (it may still exist in some places) that calculus reform meant using technology (calculators and/or computers) in calculus. In fact, rapid advances in affordable technologies have provided a powerful stimulus for rethinking mathematics curricula. But some reform projects (e.g., those at New Mexico State University and Ithaca College) started out with no emphasis on technology, and others have found that sophisticated technology alone does not assure any lasting or substantial reform. The emerging consensus recognizes the importance of using "appropriate technology," which means different things on different campuses and for different groups of students, depending on available resources, the particular focus of the course, and many other factors.

The use of technology is not just trendy, and its appearance in reformed courses is not artificial—as in, "We have to use it because it's there." In fact, computers and calculators are the tools that enable us to break out of the mindless symbol-pushing mode and allow students to experience the interplays of graphical, numerical, and symbolic representations and of discrete and continuous models and methods. Furthermore, computer-based word processing, especially with symbolic and graphic capabilities, provides an environment in which students can write and rewrite reports in which they synthesize and explain their conceptual understanding—without the heavy overhead imposed by handwritten or typewritten documents.

Finally, while many are looking at the content of calculus as what we had always hoped the traditional course would become, we are seeing some very real changes in the syllabus. First, there is now a consensus that there need not be just one syllabus—indeed, that the course will be healthier if there is more than one. Second, there is recognition that the course must not be measured out in 50-minute sound bites and 4-minute exercises.

Beyond those consensus aspects, we see a lot of different things happening, but with a number of common themes. For example, many are discovering the value of real-world problems, not as afterthoughts, but as upfront motivators, as "hooks" to capture the interest of students who have (or think they have) no interest in mathematics for its own sake. (That describes practically all students enrolled in calculus courses.) After all, one doesn't have to know much mathematics to state a substantial and interesting problem, and the problem itself can keep students focused on the task of developing mathematical tools. Here is an example from Keith Stroyan's project at the University of Iowa: Why is it that we can eradicate diseases such as polio and smallpox but not diseases such as measles and rubella?

One of the issues on which there is no consensus at all is the role (if any) of Advanced Placement (AP) calculus in a changing curricular environment. John Kenelly (Clemson University) argues elsewhere in this volume that AP as an institution can be a positive force for change. Many of the conference participants disputed that and saw AP more as a part of the problem than as a part of the solution. No one disputes that a very large number of communities all across the country have an enormous stake in AP—a stake with political, social, economic, and emotional dimensions that sometimes outweigh educational considerations.

Bill Medigovich (Redwood High School, California) suggests that universities—the folks who created AP in their own image—take responsibility for high schools (and, by implication, community colleges) in their own areas by supporting reformed high school (and transfer) calculus courses for which they will grant credit and placement, whether they are called "AP" or something else. An example: Jim Hurley (University of Connecticut) has an NSF-supported project to disseminate his technology-based calculus course to over 100 high schools across Connecticut, with credit granted by the University.

Environments for learning

"Telling is not teaching, and listening is not learning" (Sally Berenson, North Carolina State University). Every reformer either knows already or quickly discovers the need to focus on learning and on creating an environment in which learning can take place. For some, the physical location is the traditional classroom; for some, a laboratory of the "lab science" type; for some, both. In most projects, "lecture" is being partially or totally replaced by multiple classroom and/or lab activities. The operative word here is "activity"—active involvement of the student learners replacing passive reception of "the word" from an all-knowing lecturer. These classroom and lab activities often involve experimentation, discovery, and open-ended problems.

Multiple activities go hand-in-hand with multiple representations of the subject matter. Well-designed projects challenge students to use all the tools at hand—pencil-and-paper, calculator (with or without graphics), computer (definitely including graphics), and, *most importantly,* prior experience with mathematical concepts and techniques—to attack problems not previously solved, either by the students themselves or (in examples) by the instructor or textbook author. These activities are also the vehicles for written and oral communication. Students talk to each other and to the instructor about mathematics, formally (as in presentations at the blackboard) and informally (as in small group activities in or out of class or in help sessions). Students are *expected* to read their textbooks—often a new and unfamiliar expectation. And many of the activities lead much more naturally to written responses (reports or journals, for example) than to tests.

In most cases, student activities involve teamwork. Students can learn a lot from and with each other that they would never learn working in isolation or listening to a lecture. However, cooperative learning environments also raise issues on which we do not yet see consensus emerging: issues of individual versus corporate responsibility. Our students have been programmed to compete with each other, to see other students (rather than ignorance) as the enemy. And we are conditioned to using that sense of competition as a motivator and to seeing cooperation as "cheating." So how do we get students to share their learning experiences? And how do we evaluate their efforts when they do?

These are difficult questions to answer, but they are not impossible. Neil Davidson (University of Maryland) has been using cooperative learning environments in calculus and precalculus instruction for over 20 years, and he has assembled practical advice on how to do it from a number of practitioners [4]. Bill Medigovich observes that just saying "groups are good" is not enough—in particular, it doesn't work to order students to work together outside of class and continue the traditional student-to-teacher focus in the classroom. The behavior we want students to engage in has to be modeled and practiced in the classroom—under supervision—or it will never be learned. Janet Ray (Seattle Central Community College) adds that the hardest part of teaching a reformed course is to learn to structure and manage the classroom for effective learning.

Ray Marshall (former Secretary of Labor, now in economics at the University of Texas-Austin), in an address to an NSF invitational conference on systemic reform in education, made a distinction between incentives that are positive, negative, and perverse. The ones that don't work well, in economics or in education, are negative ("You do what I want, or I'll punish you") or perverse ("You do what I want, and I'll punish you anyway"). Superior achievement, both in school and in the workplace, has been demonstrated with positive incentive systems: You do what we all want, and we'll all be rewarded. In the calculus reform movement, a positive incentive system that rewards cooperative achievement has a catchy description coined by Robert White, President of the National Academy of Engineering, in an address to the 1987 NSF conference on calculus reform [8]: In the science/mathematics/engineering pipeline, calculus should become a *pump,* not a filter.

The filtering issue was a major stimulus to the reform of calculus instruction, because over half of the students taking calculus across the country were failing to achieve satisfactory grades. (Another major stimulus was the observation that, even with satisfactory grades, most students were not able to use calculus in any other course.) Consider this: Our salaries are paid by our students (or their families) through tuition and taxes. We are not paid by the medical, engineering, and business schools who want us to screen their applicants. Thus, we must assign the highest priority to *learning* (our students' best interests, whether they know it or not) and relegate *certification* to a distinctly inferior role. An important precursor to certification is assessment of student knowledge, but assessment and feedback are also important tools for education, and we should see them primarily in that light. We are responsible, first and foremost, to our students.

What we know about learning and what we are learning about learning tell us that teaching to tests is counterproductive. And now that we have a wide range of other student activities to observe—lab reports, project reports, worksheets, group interactions—we are free to de-emphasize tests as the primary focus of the course. While we may never see a consensus on just how important tests are, everyone understands that we can and should construct and use other forms of assessment as well.

Here is an example of how de-emphasis of tests can be carried out: The calculus committee at Duke, which sets policy for the laboratory course now being taken by all students who start with Calculus I, has allocated 35% of the grade to the laboratory, 20% to the final exam, and the rest to the discretion of the instructor (within broad guidelines); the instructor may use in-class tests, group or individual project reports, quizzes, homework, end-of-term essays, portfolios, or whatever he or she considers appropriate. In most sections, the split between grades given to individuals and grades given to groups is about 50–50, plus or minus 5%.

The dominant paradigm in the educational literature is constructivism, which may be described briefly as the belief that all learning is constructed by the learner in response to challenges to refine or revise what is already "known" in order to cope with new situations. This paradigm specifically contradicts the dominant belief among college-level teachers of mathematics (and probably other subjects) that knowledge is "transmitted" from knower to learner. The theory and practice of constructivism, along with the data supporting its effectiveness, are largely unknown to the mathematics faculty, but every reformer who has focused on what students learn, rather than on what teachers teach, has independently discovered at least the basic ideas of constructivism. There is a simple, highly convincing proof that constructivism better explains what's in students' heads than does "transmissionism": Ask your students to write down what they are thinking as they work through a problem. You will find in their writing ideas or thought patterns that they didn't learn from any lecture or any textbook—ideas that you wouldn't want to take credit for. Since those ideas didn't come from the external sources, their owners must have made them up.

Students

Instructors in reformed courses across the country are hearing the same complaints from their students: "This isn't what we did in high school." "You're making us work too hard." While students intend these remarks as criticisms, we see them as high praise. In courses that emphasize conceptual understanding and de-emphasize symbol manipulation, students complain that they are not learning what they need for engineering or physics or the next math course. To the extent that the rest of the curriculum remains unreformed, there may be some truth in that, but that merely tells us the job is not finished yet—not that it can't be done. Of course, it was the engineers and physicists and mathematicians teaching higher level courses who let us know that the traditional calculus course was not preparing students for their courses. The same folks, along with biologists, chemists, economists, and others, are applauding our new focus on problem solving, modeling, concepts, exposition, and technology. (As Larry Evans, chair of the Duke physics department put it, "There's nowhere to go but up.") But it takes time for this message to filter into student consciousness.

Deb Hughes Hallett has spoken often of how scary it is for students to be thrust into a totally new environment with totally new demands, when they were secure in the knowledge of what "math" is and how it works—even if they saw themselves as people who could never "do math" very well. Once we get them through that scary period—which may vary from about four weeks to a semester and a half—many of our students find that they are having fun, and we see them doing things that calculus students never did before, such as writing intelligent papers about solutions of real problems. Furthermore, they are finding meaning in the modeling and interpretation phases of problem solving, contradicting their prior beliefs that math has nothing to do with anything else and that it has no intrinsic meaning either. Janet Ray notes the new phenomena—which may occur more frequently in community colleges than in elite universities—of grades rising from semester to semester and of students taking more mathematics courses "for fun."

Most of us did not start out consciously intending to solve problems of access to mathematics and science by under-represented minorities and women. However, we have stumbled onto some solutions to those problems as well. By downplaying the roles of symbol manipulation and of traditional testing, we have created courses in which students with weak algebra skills (disproportionately, female and/or minority students) can achieve success in other ways. In so doing, they build confidence in their thinking abilities, and they often learn from their peers the necessary symbol skills they were lacking. In traditional courses, a grade of C or D is a clear indicator of worse grades to come, eventually leading to failure. In a reformed course, students can and do get stronger over time. It is no accident that courses developed by "elite" institutions such as Harvard and Duke are finding widespread acceptance at schools with large non-traditional student populations.

A key factor in expanding opportunities for women and minorities is small group work that is required, structured, and supervised. The work of Uri Treisman (University of Texas) and others has shown the importance of minority students working in groups to learn from each other and to support each other. Various reform groups are finding that this works well whether or not minority students are the majority in their groups. We are also finding, in the language of Belenky et al. [1], that women working in small groups "find their own voice" in ways that are not likely to happen in a classroom-size group. In time, as they discover that they have good ideas and can express them, women students often become vocal leaders in the larger group as well. As teachers, we have all known for a long time that the students who speak up most in class are likely to be among those who achieve the most success. Now, by starting with small-group activities, we are finding ways to make that success-through-participation accessible to almost all students.

The other side of this coin is that we have frustrated some students (disproportionately, white male students) who have been very successful at symbol manipulation in high school, and who really believe that (a) this is what mathematics is all about and (b) they want to major in mathematics. Typical question: "I can do the math—why do I have to write about it too?" Answer: "The 'math' you can do so well can be done by a machine. If you don't understand what you are doing and why you are doing it, you're not going to be a very good math major, let alone mathematician." Another typical question: "I can solve these problems by myself—why do I have to be saddled with a partner?" Answer: "Real problems are too big to be solved by one person working in isolation. That partner you may be carrying algebraically may also be carrying you conceptually. An important part of learning is learning to work with others, to bring out each person's strengths and contributions, so that the final product is greater than the sum of the parts." Many of our hot-shot symbol pushers eventually buy into this, but some do not. I am not disappointed when the latter students, frustrated by B's when they think they should be getting A's, announce that they are never going to take another math course. Better they should be frustrated as freshmen, when there is still time for them to find something they are really good at (and at which they haven't been displaced by a machine), than to have them major in mathematics and frustrate themselves and us as juniors, still rejecting the idea of "understanding" as part of mathematics.

Closely related to the problems of understanding mathematics is a fundamental problem we overlook at our peril: Most of our students are *functionally illiterate,* at least in mathematics. That is, they don't know how to read, and they don't know how to write. It is one of the great scandals of our time that students can actually "succeed" in an academic endeavor (I'm not talking about social promotion in school) without being able to read or write. We did that to ourselves and to them: We created courses in which writing was not expected, and we created "textbooks" for which reading was unnecessary. Now, as we focus on understanding, we are confronted with the folly of our past practice. Most of our institutions have discovered the concept of "Writing Across the Curriculum"—and have implemented some form of writing program, with varying degrees of success. But where are we going to find "Reading Across the Curriculum?" Do we have to invent it ourselves? If we don't, how can we expect students to actually read books about mathematics when they have never had to do that before?

As reformed courses become the "mainstream," no longer "experimental" sections that students can avoid by taking some other track, we can expect to see increased levels of frustration, bitterness, even hostility. That was already evident a year ago when Elizabeth Culotta wrote about Project CALC [3], and the situation has gotten worse at Duke, now that every student who starts with Calculus I is in a laboratory section. Hostility levels seem to be lower at schools that are not as selective as Duke or Harvard (where the students are absolutely certain that they know better than the faculty what a calculus course is supposed to be), but this is likely to become a real issue at every school that attempts serious reform.

There's an important message here for department chairs and deans about rating and rewarding the teaching effectiveness of the faculty. In a time of transition, teaching must not be evaluated solely by students—who always have a vested interest in the status quo—and it may even be necessary to consider negative evaluations as a positive sign and vice versa. Of course, that means numerical scores on five-point scales are not enough. One must listen to or read student comments very carefully to know whether their complaints are about positive or negative aspects of teaching. Better yet, samples of student reports should become part of the process of deciding whether students are in fact learning—which is the most important issue in deciding whether the faculty are in fact teaching. This is a *tough* issue: Several conference participants noted the relatively high median age of those in attendance, reflecting the risk for young faculty members in "getting involved" with reform. David Lomen (University of Arizona) observed that, to get that median age down, we have to learn to value teaching in ways other than student evaluation forms.

At Duke, we have evidence that offsets the negative attitudes displayed by our students at the end of Calculus I (and sometimes later): Jack Bookman, evaluator for Project CALC, has conducted a series of tests with sophomores and juniors, half of each group having had the Project CALC course and half the traditional course in their freshman year. Once they have had time to reflect, the same students who wrote very negative course evaluations about Project CALC have much more positive attitudes than their peers toward mathematics in general and its role in their lives [2].

Instructors

We're just like the students: Changing the way we have always "done math" in the classroom is scary. But once we get past that, we are having fun, working harder than we ever did before, and finding meaning in what we are doing. As Janet Ray notes, more faculty are talking to each other than ever before—both about mathematics and about learning. Where faculty are aware of the damage being done by traditional teaching methods and are open to possibilities for change, the idea spreads rapidly.

One of the critical ways in which we are like our students is in our avoidance of responsibility—and even those who proudly wear the "reform" label have problems with responsibility on occasion. Our students believe *we* are responsible for "educating" them; we *know* that one can educate oneself but not another. Yet, we toss the responsibility back to them without necessarily taking responsibility for providing what they need for effective learning. In particular, we resist the idea that it is our job (if nobody has done it yet) to help our students learn to read a mathematics book or to write coherent sentences and paragraphs about mathematics. It's *our* job (if nobody has done it yet) to help students learn how to work together to achieve a common goal. The list of important new jobs is much longer, but before we can begin on those, we have to educate ourselves about how to do these things. And then we have to take responsibility for helping our colleagues educate themselves as well. Frank Demana (Ohio State University) says that the biggest mistake they made in their projects was to underestimate the importance and difficulty of educating the faculty for real change.

I am often asked how we convinced our mathematics department to accept the laboratory approach as the only way calculus would be taught at Duke. (We're only half way to that "only way"—the other half of our calculus students start beyond Calculus I, and most of them are not in Project CALC sections—but the principle has been accepted.) In fact, the Duke faculty convinced themselves by volunteering to teach the laboratory course. Many of them did so under terrible conditions, when we were still experimenting with the workload and other aspects of the course. All had to cope with the problems of becoming novices again; in addition to the damage that does to one's ego, they re-learned that anything you do for the first time is a lot more work than is repetition of a comparable task. Most of our faculty experienced the rejection of negative student evaluations, even when there was no hint of any problem throughout a semester of cordial collegiality with students who were achieving at a high level. Nevertheless, almost every individual who taught a section of the course decided that, even though there were still things that needed work, we were headed in the right direction.

How much work does it take to teach a reformed course? Too much—so far. Most of the instructors I have asked this question are working much too hard to justify the effort for one course. But most are also doing the course for the first time. Those of us who have taught such courses two or more times after achieving stability of materials and syllabus have come to the conclusion that the effort required in the steady state is about the same as that required for a junior-senior level majors course. And that's exactly as it should be—our freshman and sophomores are just as important as our majors, no more and no less.

The dominant metaphor for what calculus reformers are doing in the classroom is "coaching." Our students can understand that no one learns to play basketball or violin by listening to lectures about dribbling or bowing. They can also understand that people who lack the talent to be a varsity athlete or a concertmaster can nevertheless learn to play well enough to achieve personal satisfaction. To do so, they have to start playing—preferably under the guidance and direction of a good coach. (Violin coaches are usually called "teachers"—there's probably a message there.) Furthermore, most athletic and artistic endeavors are more satisfying when practiced with other people, not in isolation. Why should intellectual endeavors be any different?

In [7] I wrote of a personal metaphor for my role in helping students through the scariness of giving up "math" and learning to do mathematics: that of hospice nurse (my wife's profession). I frequently see in my Calculus I students the classical stages of dying described by Elizabeth Kübler-Ross: denial, anger, bargaining, depression, and finally acceptance. Not everyone goes through all the stages or in the same order, and some never make it to acceptance—at least, not before the course is over and the evaluation form filled out. But this pattern is what I see in my students, and an important part of my job is to nurse them through the stages of dying in their anti-intellectual life, easing their pain and respecting their dignity, in the hope of their reaching joyful acceptance and new intellectual life.

We're just like the students: As more and more faculty and graduate students become involved in teaching reformed courses, whether by choice or because of departmental decisions, we see more and more who are not ready to give up their investment in the status quo. We must recognize that there are pains to be eased and dignities to be respected among our colleagues, just as among our students. Unfortunately, it is not clear who will be the nurses in a highly competitive academic environment that traditionally has little respect for time and effort devoted to making sure that students learn. Supportive and sensitive deans and department chairs can play the role of nurse, but most administrators were not chosen for those qualities. With or without nurses, there is an important role for "support groups"—groups of three or four faculty and graduate students who share responsibilities for a given course and who meet regularly (perhaps over lunch) to share experiences, horror stories, and solutions. Notice that this is exactly what we are recommending to, even requiring of, our students:

We're just like the students.

We're not at all like our students: Simple arithmetic tells us that the percentage of potential mathematics majors among the 600,000 or more students starting calculus each year is vanishingly small—never mind the percentage of potential mathematicians and mathematics educators. So, if we ask ourselves how we learned mathematics—that is, if we use ourselves as model learners—we're going to get it wrong. Even if every student came to us fully prepared to study calculus, they would not be like us. The fact that we learned in a system dominated by lectures—and did wonderfully well—does not prove that that system "works." Indeed, those of us who "made it" know that we didn't walk out of the lecture hall knowing more mathematics. Rather, we learned there what we had to go home and work on so it would become part of our repertoire, so we could understand the concepts and master the techniques. Most of the students we teach today have been taught, both overtly and subtly, not to do what we did to learn mathematics. They use textbooks that don't have to be read, and they learn "algorithms" for mimicking steps in worked-out examples so they can "do" homework and tests. It's widely considered "unfair" to ask students on a test to do a problem they have not done already. Our students, at least as they come to us, are not at all like us.

Nevertheless, they are our students, and we are responsible for seeing that they learn: "Teach the students you have, not the ones you wish you had" [6]. In time we can expect the preparation of college students to change for the better. As the NCTM *Standards* take hold and become part of a nationwide reform effort, many of the problems we see in getting lower division students to understand mathematics will either be dealt with in elementary and secondary schools or not be created in the first place. But we can't wait around for that change and still be responsible to the students we have now. Nor can we declare the problem unsolvable by blaming the victims—among whom I include not only the students, but also their teachers and their parents. We now have constructive proof from a large number of reform projects, being replicated on a much larger number of campuses, that we can reach students where they are. In addition to confronting them with appropriate challenges to their imperfect understanding of mathematics, we can teach them to read and write—or support the efforts elsewhere on our campuses to do that. We can gently and supportively turn students around, thereby encouraging them to become thinking and working adults, actively engaged in constructing their own knowledge of mathematics and of its power as a tool in whatever intellectual discipline they choose to pursue.

Summary and Challenges

At the urging of the mathematical community, the National Science Foundation began its efforts in undergraduate curriculum reform with calculus, which was seen to be the capstone course for secondary mathematics and the keystone course for collegiate mathematics. This strategy for eventually influencing all of the secondary and tertiary curricula in mathematics is working. Calculus reform is a viable movement that has turned out to be highly consistent with the NCTM *Standards,* and therefore with the emerging reform of school curricula. Challenge: Our *self-interest* as college faculty demands our support for reform at the elementary and secondary levels—the best solution in the long run to the problems of functionally illiterate and scared students.

Reformed calculus courses are leaner in terms of the number of topics in the syllabus, but much richer in actual content. They have been enriched by balanced emphases of symbolic, numeric, and graphical representations, by reading, writing, and speaking as learning and assessment tools, by use of appropriate technologies (calculators and computers), and by extended and realistic problems. Challenges: We must start now to reform our own post-calculus courses, which are not yet taking advantage of the newly-developed abilities of our calculus students. We must build bridges to client departments, both to make sure they understand what we're doing (and why) and to influence change in their courses that require calculus.

Calculus reform has come to mean reform of pedagogical methods as much as reform of the curriculum. Most reformers are beginning to pay attention to the literature on learning, which contains a strong constructivist message: Listening to lectures is not an effective way to learn—but active involvement, especially in small group activities, is. The same literature also tells us that we can set high expectations for our students, and, with proper support, they will achieve those objectives. Finally, this literature tells us of the importance of posing problems that are interesting and meaningful to our students at their current stage of development—not problems that would be interesting or meaningful only if they had already achieved some higher level of mathematical sophistication. Challenges: We must reorient ourselves and our colleagues toward teaching and learning. To support our need and our students' need for two-way written communication, we must buy into "Writing Across the Curriculum," and we may have to invent "Reading Across the Curriculum."

Technology, ranging from inexpensive scientific calculators to powerful symbol manipulation systems, is playing an important role in shaping both content and methodology—content because the existence of technology alters the relative importance of specific techniques, and methodology because technology offers opportunities for creating new learning environments (e.g., laboratories). Challenge: Mathematics courses that ignore technology perpetrate a fraud—*we have to educate our students for the world they will live in, not the one we grew up in.*

Now that many of us are beyond the first flush of excited volunteers, we are finding a lot of resistance and negative reaction from students who don't want us to shake up their comfortable relationship with "math," no matter how distasteful that relationship may be. Challenge: We (and our administrators) must be careful not to let students be the sole evaluators of our efforts, especially when students lack the perspective from which to sensibly evaluate what they have learned. Rather, we (and our administrators) must look at what students actually do, both in our courses and beyond.

Reform in calculus instruction places fundamentally new demands on us as teachers. We have to be open to new ways of relating to students in the classroom—indeed, be willing to become novices again. Administrators have to invest new resources in developing the teaching faculty, providing for an appropriate faculty-student ratio, and supporting workshops and other means of upgrading skills. Furthermore, we have to devote more of ourselves to teaching as a profession, engaging in dialogues on matters of pedagogy, forming support groups, and—perhaps most important—valuing the profession for which we get paid. The rewards are worth it: We may not ever agree that we get paid what we are worth, but nothing quite matches the thrill of seeing students' eyes light up when they really understand.

There is good news and bad news, both from that great philosopher of the comic strips, Pogo. The bad news: "We have met the enemy, and he is us." The good news: "We are faced with insurmountable opportunities."

References

1. Belenky, M. F., et al., *Women's Ways of Knowing*, Basic Books, 1986.

2. Bookman, Jack, "Evaluation Update," *The Project CALC Newsletter* **5,1** (Fall, 1992), 4–5.

3. Culotta, Elizabeth, "The Calculus of Education Reform," *Science* **255** (1992), 1060–1062.

4. Davidson, Neil (ed.), *Cooperative Learning in Mathematics: A Handbook for Teachers*, Addison-Wesley, 1990.

5. National Council of Teachers of Mathematics, *Curriculum and Evaluation Standards for School Mathematics*, NCTM, 1989.

6. National Research Council, *Moving Beyond Myths: Revitalizing Undergraduate Mathematics*, National Academy Press, 1991.

7. Smith, D. A., "Opinion: What's Better About Reformed Calculus," *UME Trends* **3**, Number 2 (May, 1991), 5–6.

8. Steen, Lynn A., ed., *Calculus for a New Century: A Pump, Not a Filter*, MAA Notes No. 8, 1987.

9. Tucker, Thomas W., ed., *Priming the Calculus Pump: Innovations and Resources*, MAA Notes No. 17, 1990.

Bibliography

Calculus Reform

Culotta, Elizabeth, "The Calculus of Education Reform," *Science* **255** (1992), 1060–1062.

Douglas, Ronald G., ed., *Toward a Lean and Lively Calculus*, MAA Notes No. 6, 1986.

Flashman, Martin, "Editorial: A Sensible Calculus," *The UMAP Journal* **11** (1990), 93–95.

Goodman, "Toward a Pump, Not a Filter," *Mosaic* **22** (Summer 1991), 12–21.

Smith, D. A., and L. C. Moore, "Duke University: Project CALC," pp. 51-74 in T. W. Tucker (ed.), *Priming the Calculus Pump: Innovations and Resources*, MAA Notes No. 17, 1990

Steen, Lynn A., ed., *Calculus for a New Century: A Pump, Not a Filter*, MAA Notes No. 8, 1987.

Tucker, Thomas W., ed., *Priming the Calculus Pump: Innovations and Resources*, MAA Notes No. 17, 1990.

Cooperative and Collaborative Learning; Learning Communities

Davidson, Neil (ed.), *Cooperative Learning in Mathematics: A Handbook for Teachers*, Addison-Wesley, 1990.

Finkel, D. L., and G. S. Monk, "Teachers and Learning Groups: Dissolution of the Atlas Complex," pp. 83–97 in *Learning in Groups*, Jossey-Bass, 1983.

Gabelnick, F., J. MacGregor, R. S. Matthews, and B. L. Smith, *Learning Communities: Creating Connections Among Students, Faculty, and Disciplines*, Jossey-Bass, 1990.

Heller, Patricia, Ronald Keith, and Scott Anderson, "Teaching problem solving through cooperative grouping. Part 1: Group versus individual problem solving." *American Journal of Physics* **60** (7), July 1992, 627–636.

Heller, Patricia, and Mark Hollabaugh, "Teaching problem solving through cooperative grouping. Part 2: Designing problems and structuring groups." *American Journal of Physics* **60** (7), July 1992, 637–644.

Johnson, D. W., R. T. Johnson, and K. A. Smith, *Cooperative Learning: Increasing College Faculty Instructional Productivity*, ASHE-ERIC Higher Education Report No. 4, The George Washington University, 1991.

MacGregor, J., "Collaborative Learning: Shared Inquiry as a Process of Reform," pp. 19–30 in *The Changing Face of College Teaching* (M. D. Svinicki, ed.), Jossey-Bass, 1990.

What Works, Vol. I: Building Natural Science Communities: A Plan for Strengthening Undergraduate Science and Mathematics, Project Kaleidoscope, 1991.

The State of Mathematics Education

Leitzel, J. R. C., ed., *A Call for Change: Recommendations for the Mathematical Preparation of Teachers of Mathematics*, MAA Report, 1991.

Madison, B. L., and T. A. Hart, *A Challenge of Numbers: People in the Mathematical Sciences*, National Academy Press, 1990.

Mathematical Sciences Education Board, *Counting on You: Actions Supporting Mathematics Teaching Standards*, National Academy Press, 1991.

National Council of Teachers of Mathematics, *Curriculum and Evaluation Standards for School Mathematics*, NCTM, 1989.

National Research Council, *Everybody Counts: A Report to the Nation of the Future of Mathematics Education*, National Academy Press, 1989.

National Research Council, *Renewing U. S. Mathematics: A Plan for the 1990's*, National Academy Press, 1990.

National Research Council, *Moving Beyond Myths: Revitalizing Undergraduate Mathematics*, National Academy Press, 1991.

Schwartz, J. L., "The Intellectual Costs of Secrecy in Mathematics Assessment," pp. 132-141 in *Expanding Student Assessment* (Vito Perrone, ed.), ASCD, 1991.

Sigma Xi, *Entry-Level Undergraduate Courses in Science, Mathematics, and Engineering: An Investment in Human Resources*, 1990.

Steen, Lynn A., ed., *Heeding the Call for Change: Suggestions for Curricular Action*, MAA Notes No. 22, 1992.

Teaching and Learning: General

About Teaching (newsletter), The Center for Teaching Effectiveness, University of Delaware.

Alverno Magazine, May 1992, issue devoted to assessment and outcome-based learning, Alverno College, Milwaukee, WI.

Belenky, M. F., et al., *Women's Ways of Knowing*, Basic Books, 1986.

"Bridging the Gap Between Education Research and College Teaching," NCRIPTAL, Accent Series No. 9, 1990.

Finster, David, "Applying Development Theory May Improve Teaching," *Wittenberg Today*, Wittenberg College, 1987.

Glasersfeld, Ernst von, "Cognition, Construction of Knowledge, and Teaching," *Synthèse* 80 (1989), 121–140.

Halloun, I. A., and David Hestenes, "The Initial Knowledge State of College Physics Students," *American Journal of Physics* **53** (1985), 1043–1055.

Kozma, Robert B., and Jerome Johnston, "The Technological Revolution Comes to the Classroom," *Change* **23** (Jan/Feb 1991), 10–23.

National Center on Postsecondary Teaching, Learning, & Assessment Newsletter, Pennsylvania State University, 1992.

National Teaching & Learning Forum, The (newsletter), George Washington University, 1991.

Reedy, George, "I am not at all convinced that we professors 'educate' students. What we do is force them to use their minds." *The Chronicle of Higher Education*, Dec. 19, 1990, p. B5.

"Personal Growth as a Faculty Goal for Students," NCRIPTAL Accent Series No. 10, 1990.

Steen, Lynn A., "Out from Underachievement," *Issues in Science and Technology*, Fall 1988, 88–93.

Steen, Lynn A., "Reaching for Science Literacy," *Change*, July/August 1991, 11–19.

Taylor, Kathe, and Bill Moore, "Who's Making the Meaning in the Classroom: Implications of the Perry Scheme," POD Conference paper, 1984.

Tobias, Sheila, *They're Not Dumb, They're Different: Stalking the Second Tier*, Research Corporation, 1990.

Tobias, Sheila, *Revitalizing Undergraduate Science: Why Some Things Work and Most Don't*, Research Corporation, 1992.

"Teaching Thinking in College," NCRIPTAL Accent Series No. 7, 1990.

"What Are Academic Administrators Doing to Improve Undergraduate Education?," NCRIPTAL Accent Series No. 8, 1990.

What Works, Vol. II: Resources for Reform, Project Kaleidoscope, 1992.

Teaching and Learning: Mathematics

Ball, Deborah L., "Unlearning to Teach Mathematics," *For the Learning of Mathematics* 8 (1988), 40–48.

Benezet, L. P., "The Teaching of Arithmetic I, II, III: The Story of an Experiment," *Humanistic Mathematics Newsletter* **6**, May 1991, pp. 2–14 (reprinted from *The Journal of the National Education Association*, Nov. 1935, Dec. 1935, Jan. 1936).

Borasi, Rafaella, "The Invisible Hand Operating in Mathematics Instruction: Students' Conceptions and Expectations," *NCTM Yearbook 1990*, pp. 174–181.

Bullock, Richard, and Richard Millman, "Mathematicians' Concepts of Audience in Mathematics Textbook Writing," *PRIMUS* **2** (1992), 335–347.

Cipra, Barry A., "Untying the Mind's Knot," in *Heeding the Call for Change* (L. A. Steen, ed.), MAA Notes No. 22, 1992, pp. 163–181.

Cobb, George, "Teaching Statistics: More Data, Less Lecturing," *UME Trends*, October 1991, pp. 3, 7.

Kulm, Gerald, ed., *Assessing Higher Order Thinking in Mathematics*, American Association for the Advancement of Science, 1990.

MER Newsletter, Mathematicians and Education Reform Network, University of Minnesota, 1989.

Mitchell, Richard, "The Preconception-Based Learning Cycle: An Alternative to the Traditional Lecture Method of Instruction,"*PRIMUS* **2** (1992), 317–334.

Monk, G. S., "Students' Understanding of Functions in Calculus Courses," *Humanistic Mathematics Network Newsletter*, No. 2, March 1988.

Papert, Seymour, *Mindstorms: Children, Computers, and Powerful Ideas*, Basic Books, 1980.

Pólya, George, "Reprints of Papers on Teaching and Learning in Mathematics," pages 473–603 in *George Pólya: Collected Papers*, Vol. IV, edited by Gian-Carlo Rota, MIT Press, 1984.

Selden, Annie, and John Selden, "Constructivism in Mathematics Education: A View of How People Learn," *UME Trends*, March 1990, p. 8.

Selden, John, Alice Mason, and Annie Selden, "Can Average Calculus Students Solve Nonroutine Problems?" *Journal of Mathematical Behavior* **8** (1989), 45–50.

Selden, John, Annie Selden, and Alice Mason, "Even Good Calculus Students Can't Solve Nonroutine Problems," preprint, 1990.

Schoenfeld, Alan H., "When Good Teaching Leads to Bad Results: The Disasters of 'Well-Taught' Mathematics Courses," *Educational Psychologist* **23** (1988), 145–166.

Zimmermann, Walter, and Steve Cunningham, eds., *Visualization in Teaching and Learning Mathematics*, MAA Notes No. 19, 1991.

Technology in Teaching and Learning

"The Computer Revolution in Teaching," NCRIPTAL Accent Series No. 5, 1989

Johnston, J., and S. Gardner, *The Electronic Classroom in Higher Education: A Case for Change*, NCRIPTAL, 1989

Leinbach, Carl, ed., *The Laboratory Approach to Teaching Calculus*, MAA Notes No. 20, 1991.

Smith, David A., et al., eds., *Computers and Mathematics: The Use of Computers in Undergraduate Instruction*, MAA Notes No. 9, 1988.

Smith, D. A., and L. C. Moore, "Project CALC: An Integrated Laboratory Course," pp. 81–92 in Carl Leinbach (ed.), *The Laboratory Approach to Teaching Calculus*, MAA Notes No. 20, 1991.

Writing as a Teaching and Learning Tool

Bell, Elizabeth, and Ronald Bell, "Writing and Mathematical Problem Solving: Arguments in Favor of Synthesis," *School Science and Mathematics* **85** (1985), 210–221.

Connolly, Paul, and Teresa Vilardi, eds., *Writing to Learn Mathematics and Science*, Teachers College Press, 1989.

Gopen, G. D., and D. A. Smith, "What's an Assignment Like You Doing in a Course Like This?: Writing to Learn Mathematics," *The College Mathematics Journal* **21** (1990), 2–17. Reprinted from Connolly and Vilardi cited *supra*.

Sterrett, Andrew, ed., *Using Writing to Teach Mathematics*, MAA Notes No. 16, 1990.

College Precalculus Courses: Challenges and Changes

Franklin Demana
Ohio State University

The number of funded projects designed to reform college precalculus mathematics pales when compared with the number of funded calculus reform projects. Some of this can be attributed to the fact that such courses are often not considered college level courses so they attract very little faculty interest. I give an overview of a few of these precalculus projects and describe some common themes in content and in instructional principles. Several examples are given to illustrate some of the themes.

Projects

1. The United States Military Academy (David C. Arney, Frank R. Giordano)

The United States Military academy adopted a new core curriculum in the fall of 1990. The core consists of one Discrete Dynamical Systems course, two Calculus courses, and one Probability and Statistics course, taught on a semester system. Mathematics is approached symbolically (algebraically), graphically, and numerically. All of the courses use discovery and experimentation, and writing and modeling are part of the courses.

The Discrete Dynamical Systems course has a heavy emphasis on modeling using examples from the physical and social sciences. Students learn to model with difference equations and to determine solutions symbolically, graphically, and numerically. Discrete systems lend themselves to examination using iteration by a spreadsheet or CAS package. The students also learn basic matrix algebra by finding solutions of systems of equations, eigenvalues, and eigenvectors. Then they learn to solve systems of difference equations. Considerable time is spent on analysis of nonlinear equations. This course anchors their mathematics education in practical modeling experiences and is accessible from high school algebra. It allows an orderly and logical transition to the study of both calculus and differential equations.

2. Math Modeling/Precalculus Reform Project (Sheldon P. Gordon, Ben A. Fusaro)

This project involves a dramatically different alternative to the standard precalculus course. The goal is to emphasize the qualitative, geometric, and computational aspects of mathematics within a framework of mathematical modeling at a level appropriate to precalculus students. Applications drive the mathematical development. The mathematical knowledge and skills needed for calculus are introduced, developed, and reinforced in the process of applying mathematics to model and solve interesting and realistic problems.

The materials provide a one-semester course that will lay a different, but very effective, foundation for calculus, and for a one- or two-semester course that will stand as a contemporary capstone to the mathematics education of students who do not plan to take calculus.

The models developed are based primarily on difference equations, data analysis, probability, and matrix algebra. The focus on difference equations includes applications of first- and second-order difference equations and systems of first-order difference equations. The emphasis is on modeling a variety of situations and interpreting the behavior of the solutions in terms of the situations. For example, an assortment of models of growth and decay processes (for populations, diseases, technology, etc.) using both first-order equations and systems of equations are developed. The treatment of second-order linear difference equations includes a treatment of simple harmonic motion for a mass on a spring as well as the case of a system with damping. Other applications include projectile motion in the plane, effectiveness of sorting methods in computer science, and other models from physics, biology, economics, and sociology.

Data analysis techniques are also introduced early and used throughout so that students learn how to interpret data values that arise in many different contexts. For instance, regression analysis is developed, including nonlinear regression as a way of reinforcing notions on the behavior of different classes of functions. The computational drudgery is relegated to a computer or graphing calculator with statistical functions. Further, the focus is on collecting and analyzing real data sets and matching them to appropriate mathematical models and interpreting the results.

Another thread which is interwoven throughout these courses is the notion of probabilistic ideas in the context of performing random simulations. These include Monte Carlo simulations to estimate the value of π, to estimate the average value of a function on an interval or to estimate how often polynomials have complex roots, studies of radioactive decay and spread of disease, and development of models to investigate various waiting-time situations. In addition, geometric probability is emphasized as a vehicle for reinforcing geometric and trigonometric ideas in a new setting.

The course also involves a considerable degree of computer or graphing calculator work to explore the implementation of mathematical models. It involves a variety of live classroom experiments to investigate the accuracy of mathematics in predicting the results of actual processes or to help develop mathematical models based on observed experimental data. For instance, Newton's Law of Cooling, the damped spring, and Toricelli's Law on fluid leaking out of a cylinder are treated experimentally as well as theoretically.

The course also features a series of student investigations to provide a real-life dimension to the mathematics. For example, students conduct individual investigations of mathematical ideas and methods using computer software. Students collect sets of data of interest to them, say on some growth process or on the acceleration times of a Porsche, and eventually determine the best fit curve, possibly exponential, logistic or power. Similarly, they have been asked to model the number of hours of daylight in a city of their choice based on actual data collected from the newspapers or other resources.

3. Consortium for Mathematics and Its Applications Inc. (Walter Meyer, Sol Garfunkel)

This project is developing a two-semester survey course suitable for mathematics majors as well as many other majors. It is intended to serve as an alternative to calculus as a gateway course for college level mathematics. The course foreshadows much of the undergraduate curriculum for mathematics majors in the same way that an introductory survey course would in psychology, physics, or economics. By stressing modern applications they hope to give a better and more rounded picture of today's mathematics and capture student interest.

Calculus is not a prerequisite, but the course presumes some knowledge of functions, including the logarithmic and exponential functions. Students should have a strong high school background and be at the level of maturity and seriousness of purpose normally expected of those ready to enroll in calculus.

The course will have six major units.

1. Change An introductory study of difference equations and dynamical systems using applications such as population growth of single species and ecosystems, Newton's Law of Cooling, pollution control, and chaos. By using discrete methods rather than instantaneous methods, these applications need not be delayed till after the study of calculus.

2. Geometry Descartes' idea of coordinate geometry is presented in a modern guise of vector geometry, mostly in 2 dimensions. Applications are 20th and 21st century, including robotics and management science (linear programming).

3. Matrices This material is a crossroads of much of mathematics. Examples from the previous units are revisited from a more sophisticated viewpoint and new ideas are presented in a concrete way.

4. Discrete Mathematics This material is inspired by modern developments in computer science and includes: network structures, elementary combinatorics, and the Boolean logic which lies at the heart of computer chips.

5. Probability An outline of the attempt to tame Chance and Randomness using mathematics. Independence and conditional probabilities. Many applications in the social sciences are included.

6. Algebra The use of symbols to represent numbers, often "unknowns," has its counterpart in symbolic manipulations of symmetry operations on geometric objects. An elementary approach to groups and Burnside theory will be developed and applied to enumerating structures such as chemical molecules of certain classes. Error-correcting codes and other types of codes.

4. The Calculator and Computer Precalculus (C^2PC) Project (Frank Demana, Bert Waits)

The project began as an extension of earlier work at The Ohio State University that used calculators to enhance the teaching and learning of mathematics with remedial students. We have learned that change occurs in small steps. We have learned the importance of taking a familiar body of material and then assuming that computer graphing (function and parametric) would be available to all students on a regular basis for both in-class activities and homework.

The project began with desktop computers and evolved to using graphing calculators. Technology is used to enhance the teaching and learning of mathematics. The greatest benefits seem to come from technology that is interactive and under student and teacher control. Computers and graphing calculators promote student exploration and enable generalization. Computer-based technology makes graphing a fast and effective problem solving strategy. Graphical representations of problem situations are easily added to the usual algebraic representations. More realistic problem situations and deeper mathematics are possible with the aid of technology. This approach permits a focus on problem solving and encourages generalization based on strong geometric evidence. Classroom instructional models that encourage students to be active partners in the learning process are a natural consequence. Technology available today need not radically change the mathematics content that is traditional in mathematics, but use of this technology can dramatically change the way mathematics is *taught and learned*.

C^2PC teachers use two important technology-driven instructional models. Students participate in an interactive lecture-demonstration instructional model in a classroom with a single computer. Computer laboratories and classrooms where students have graphing calculators allow a guided-discovery instruction. Teachers use a carefully prepared sequence of questions and activities to help students understand or discover important mathematical concepts. Graphics calculators are used regularly on homework assignments. C^2PC students come away from their experience with strong intuition and understanding about functions, important to success in calculus, advanced mathematics, and science.

The C^2PC content is easily recognizable. However, the tools used to teach and learn the familiar content are new and different.

We approached the incorporation of hand-held visualization technology as a natural evolution of our positive experience with scientific calculators. As the project evolved, we found we used computer (graphing calculator) visualization tools to do ten types of fundamental activities that occurred repeatedly in every chapter of our project materials and every day in our project classrooms:

1. Approach problems numerically.
2. Visually support the results of applying algebraic paper and pencil manipulations to solve equations and inequalities.
3. Use visual methods to solve equations and inequalities and then confirm using analytic algebraic paper and pencil methods.
4. Model, simulate and solve problem situations and confirm, when possible using analytic algebraic paper and pencil methods.
5. Use computer generated scenarios to illustrate mathematical concepts.
6. Use visual methods to solve equations and inequalities that cannot be solve using analytic algebraic methods.
7. Conduct mathematical experiments; make and test conjectures.
8. Study and classify the behavior of different classes of functions.
9. Foreshadow concepts of calculus.
10. Investigate and explore the various connections among different representations of a problem situation.

Perhaps the most important thing we learned in addition to the principle of incremental change was that algebra has a new role in mathematics, as a language of representation, rather than as a tool for paper and pencil manipulation. It is clear to us that what is important today is the students' ability to correctly "algebraically represent" when confronted with a "real problem" rather than apply paper and pencil manipulations to a series of routine typical textbook problems. Other new ideas were "tool related" such as screen coordinates versus math coordinates.

We found that a technology-enhanced graphing approach also made some old topics such as scale and error far more important. We also had to confront many pitfalls (teaching opportunities!) such as function behavior hidden from view in a graph. For example, it is easy to miss the extrema of $y = x^3 - 2x^2 + x - 30$. (Try it!)

We learned that the teaching and learning process changes when all students use hand- held technology. We observed a significant change in our teaching behavior and the behavior of our project teachers. We now lecture much less. We see the teacher as a guide or advisor. C^2PC students actively investigate, explore, make, and test conjectures, and solve real problems. We learned the value of group or cooperative learning especially with high school seniors and college freshman. Our teachers reported that students' interest in mathematics and mathematics communication skills increased. Talking about mathematics became common!

5. A Laboratory Approach to Precalculus (Marsha Davis)

This project uses graphing technology as an integral part of a precalculus course. Its authors target concepts that have traditionally been difficult for students to grasp (piecewise functions, asymptotic behavior, mathematical modeling of real-life phenomena) and concepts that have been previously inaccessible to precalculus students (optimization, solutions to certain equations). These topics lend themselves to exploration on a computer or graphing calculator. The labs are not interchangeable units: one lab is likely to refer to a previous one, showing how to use an earlier concept in a new setting or building on a previously developed concept. The labs are collaborative, and the mathematics is presented in real-world situations. Lab worksheets are given to the students ahead of time to allow for planning. Students formalize their ideas in a written lab report.

One lab explores exponential functions. It begins with the classic "chessboard problem" to show the results of repeated doubling. Students explore the behavior of exponential functions with various bases and discover why the number e plays such an important role in exponential growth. Two different sets of data are provided to study the AIDS epidemic. In one case, the spread of the disease has not been slowed, and a straightforward exponential model can be used. In the other, the number of cases has leveled off, and the best model seems to be an exponential function joined to a horizontal line. Students are encouraged to interpret these models in terms of the actual situation.

Common Content Themes

The use of interactive graphics offers an opportunity to change emphasis in the mathematical content of precalculus. Most calculus instructors have based their instruction on the assumption that a picture or graph explains all. The assumption is that graphs provide a great deal of information about problems and their algebraic representations. Recent experience indicates that graphs become intuitive only after students have learned how to read and interpret the information they provide. That is, students must be *taught* what is contained in graphs before they can serve as a basis for intuition. Following are some of the major content emphases that are exploited to build an understanding of graphs and functions:

1. Interactive graphing Interactive computer graphing is used to provide a rich array of examples of graphs and functions for students to explore and examine. This gives the student many opportunities to form generalizations and to develop concepts about graphs, functions, and their characteristics.

2. Problems as means Real world problems are used to approach and teach concepts and skills instead of merely as exercises after concepts have been taught. Often a problem serves as the stimulus for a discussion of some new mathematics with the new mathematics serving as the conclusion for that discussion.

3. Calculus topics without calculus The mathematics is organized differently from standard texts. Many topics that are treated lightly or not at all in other precalculus texts are explored in depth. Limits, asymptotes, extrema, continuity, and other topics that foreshadow calculus are given a thorough treatment.

4. Viewing rectangles and scale With either paper-and-pencil graphing or a computer graphing utility, one only examines a portion of most graphs. A viewing rectangle specifies the portion of the plane within which a graph is to be examined; that is, the minimal and the maximal values of the x-coordinates and the y-coordinates. Students learn how to pick viewing rectangles to satisfy the purpose of the problem at hand; they learn how to zoom-in, zoom-out, and choose different scales for the two coordinate axes.

5. Local behavior of functions The ability to change viewing rectangles permits students to examine closely the graphical behavior of functions. Students can zoom-in to see at close range such local features as extrema and intercepts.

6. End or global behavior of functions Alternatively, students can zoom-out to obtain a global view of the graph. The notion of asymptote comes alive for students in a new and richer way.

7. Graphical solution algorithms Equations, inequalities, and systems of equations can be solved graphically by a zoom-in procedure. This graphical approach is powerful. Equations involving any elementary functions can be solved by the zoom-in method; whereas their algebraic solutions require a myriad of paper-and-pencil methods.

8. Parameters and functions Students can explore and discover the effects that equation parameters have on the graph of a function. For example, students can experiment with the equation $f(x) = ax^2 + bx + c$ to determine the effects that a, b, and c have on the graph of a quadratic function.

9. Mathematical modeling Students investigate a wide variety of challenging problems. They create algebraic and geometric representations for a given situation and use these representations in the solution of the associated problem.

Common Instructional Themes

One of the key factors affecting success in using a graphing utility is the arrangement of the classroom. Graphs should serve as the stimulus for discussion of the mathematics by the class. This means that students must be able to see what is being discussed. For a computer display of a graph, there are two main operating modes:

- Laboratory mode with each student at a separate computer or graphing calculator. Requiring each student to purchase a graphing calculator is usually not a problem. Some bookstores are now buying and reselling graphing calculators. Some colleges use a rent or rent with option to buy program.

- Demonstration mode with a single large monitor, with an overhead projector palette display, or with a graphing calculator overhead model.

Both modes work for instruction. Each has advantages and disadvantages. It is somewhat easier to focus attention on critical concepts in the demonstration mode because the display is limited to what is being examined. In the demonstration mode, it is convenient to have a student at the computer keyboard making the entries under your direction. This allows you to move around the room observing what students are doing while you are leading the discussion. Being freed from the entry of mathematics into the computer gives you considerably more flexibility in reacting to the class and in observing what your students are doing and thinking.

Active Involvement

Students learn mathematics best when they are actively involved. One feature of graphing utilities is that students are much more inclined to deal directly with the mathematics rather than to sit back passively and listen to a teacher's presentation. The materials can be organized to take advantage of the ability to explore modifications in the mathematics painlessly and without labor.

Suppose a student enters the functions $f(x) = 3x^2-2$, $f(x) = 3x^2 - 3$, $f(x) = 3x^2 - 12$, $f(x) = 3x^2 - 5$, $f(x) = 3x^2 + 7$, $f(x) = 3x^2 + 32$, $f(x) = 3x^2$, and $f(x) = 3x^2-3/2$. After each entry, the student examines the graph of the new function overlaid over the previous function(s). The student is naturally and directly seduced into associating the changes of constant term with the different placements of the graphs. The visual and the symbolic are intertwined actively in a fashion that was not possible when the student had to interrupt thinking to construct the graphs necessary for the comparisons.

It is easy to shift to a questioning style where you begin by saying, "*What happens* if you change the function by ...?" The act of graphing does not interpose itself between the mathematical point that you want to make and the student's being able to deal with the mathematical question. In the past, the time-consuming act of graphing forced a delay of the critical concept until the graph(s) had been produced. This forced the student to a passive examination of completed graphs rather than the student being able to construct quickly the graphs that made the comparison possible.

Students can actively pursue ideas by making minor modifications in their graphs. Further, the teacher can readily check what the student has displayed to ascertain the idea the student is encountering. In a very real sense, the ready display means the student is "on the spot" and not able to escape the consideration of the attentive teacher.

Verbal Interaction

The nature of teacher-student talk is changed by the use of the graphing utility. Indicated in part by the phrase "What happens if ..." discussed above, it is consistent with what is important in learning mathematics. Students need the opportunity to test their ideas with the teacher and with their peers. Better, more efficient learning occurs when students offer their ideas for reaction and shaping. Many ideas in mathematics are collections of subtle conditions. If the teacher only has access to the written work of the student, misconceptions are hard to identify. Talk allows the identification or diagnosis of students' misunderstandings and, furthermore, provides an opportunity for the teacher or another student to react and correct. Changes by the student in reaction to comments serve to correct the misunderstandings and change problem approaches in constructive ways.

Many students are quite sloppy in talking and thinking about graphs. For example, many have only had experience with scales that are the same on the x-axis and the y-axis. Many of the realistic problem situations do not graph in a usable, interpretable form if the same unit is used on both axes. The graphing utilities force students to deal with the problems of choosing and interpreting scale. The most painless way to sharpen language and to think about scale is to have students describe their solutions. Peers are very willing to say, "I don't understand." Students are forced to rethink the situation and communicate ideas more clearly.

Students are very sloppy early in the course in their use of the words above and below. For example, in describing the solution to the inequality $|x - 3| > 6$, they may graph $f(x) = |x - 3|$ and $g(x) = 6$ on the same coordinate system and then say the solution is all of the graph of f that is "above" the graph of g. "Above" fails to indicate that the solutions are values of x and that some are greater than nine and some are less than negative three. Indeed, listening to the students talk indicates that the connection between a graphical representation and an algebraic representation is quite weak for many students at this level.

Verbal interaction is essential to the successful teaching with technology materials for two primary reasons:

1. It gives you information about what the student is thinking.
2. It allows you to shape ideas in correct directions.

Often the comment of a student will provide a "teachable" moment that you can capitalize on to extend your instruction to ideas not directly covered in the materials. You will find that having the capability of graphing quickly and easily allows you to pursue and explore conjectures in a manner that was not possible using paper-and-pencil graphs. Graphing utilities can be used as a stimulus for thinking and discussion.

Reading

The new materials are different from many precalculus texts because students are expected to read and to experiment with the mathematics using a graphing utility. Many current texts do not expect students to read; rather the first portion of a lesson encapsulates a rule that can be applied to exercises that have little variation (and less problem solving). Students are expected to read. In the past you would have worried that your students did not "push a pencil" enough while studying; the shift inherent in these materials is for students to "push the graphing utility" to try variations on the mathematics they are reading. (And students are expected to "push a pencil" while studying!)

The typical student early in the course does not realize the importance of reading in mathematics. The student can gain a great deal of understanding about the language and logic of mathematics from extensive reading and working through the textbook's examples.

Generalization

The graphing utilities offer the teacher an opportunity to help students form generalizations about functions and graphs. Suppose, for example, that you are working with a polar coordinate graphing utility. It is easy in the course of a few button pushes to have students examine the graphs of $r = f(\theta) = \theta^1$, $r = f(\theta) = \theta^2$, $r = f(\theta) = \theta^3$, $r = f(\theta) = \theta^4$, $r = f(\theta) = \theta^5$, $r = f(\theta) = \theta^6$, $r = f(\theta) = \theta^7$, and $r = f(\theta) = \theta^8$. Then the students can describe what they see and formulate a generalization about the effects of a change of exponents on the spiral $r = f(\theta) = \theta^n$. Before graphing utilities were available, it was difficult for the teacher to organize a lesson that would readily and efficiently lead students to generalizing, since constructing the graphs was so time-consuming.

Throughout the new courses you will want to explore the power of the graphing utilities to allow students to create many different graphs to form a base for generalizing. This power in organizing a lesson is new territory for instruction.

Some Examples

Example 1 shows how paper and pencil algebra processing can be supported visually.

Example 1. Solve the inequality $x^3 - 4x < 0$.

Solution. By factoring and applying a sign pattern method, the solution is $(-\infty, -2) \cup (0, 2)$. Students make the connection that the solution of $x^3 - 4x < 0$ is also given by the values of x for which the graph of $y = x^3 - 4x$ is below the x-axis. Thus the graph in Figure 1 provides support for the analytically determined answer. The notation $[-4, 4]$ by $[-10, 10]$ below the figure describes the rectangular "viewing window" determined by $-4 \leq x \leq 4$ and $-10 \leq y \leq 10$ which is set on most graphing calculators by using the RANGE key.

This type of activity empowers students with the ability to make connections among the algebraic and geometric representations of the "problem" (in this case the

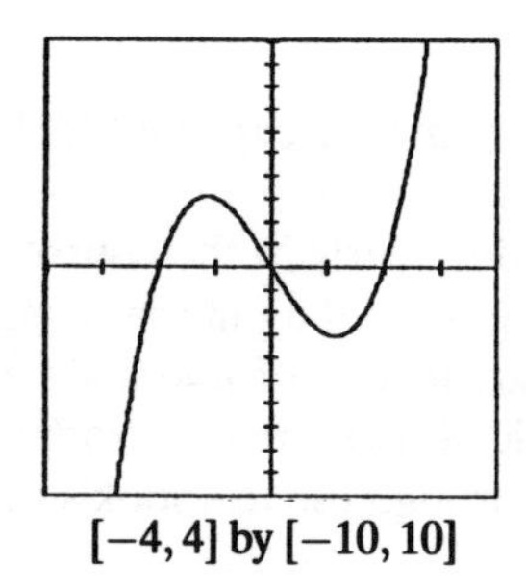

$[-4, 4]$ by $[-10, 10]$

Figure 1: A complete graph of $y = x^3 - 4x$. The graph is below the x-axis for x in the interval $(-\infty, -2) \cup (0, 2)$.

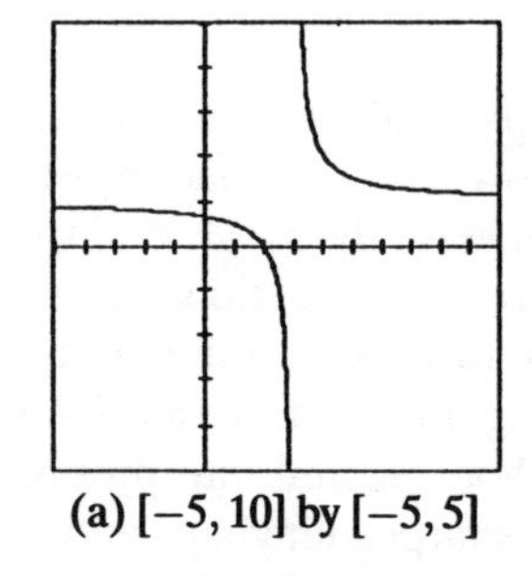

(a) $[-5, 10]$ by $[-5, 5]$

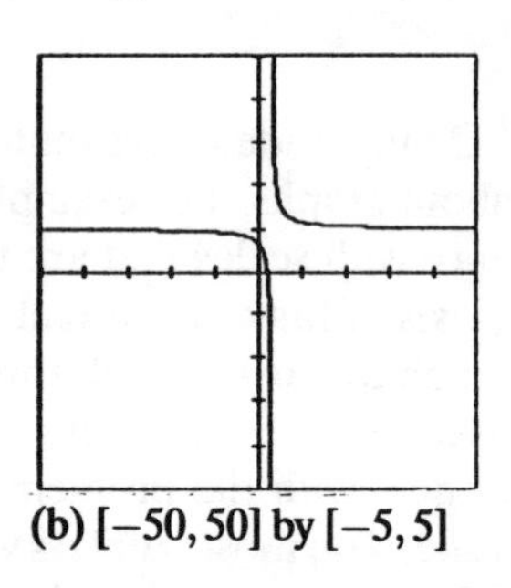

(b) $[-50, 50]$ by $[-5, 5]$

Figure 2: Two views of $y = (x-2)/(x-3)$ showing vertical and horizontal asymptotes.

inequality) and to support (check) their solutions found analytically, visually. This activity is called "do algebraically and support graphically."

The next activity shows how properties of classes of functions can be investigated using graphing technology.

Example 2. Discuss the asymptotes of rational functions of the form $y = N(x)/D(x)$, where N and D are polynomials of degree 1, 2 or 3.

Solution. Degree 1 in both numerator and denominator leads to the general form: $y = (Ax+B)/(Cx+D)$, where $C \neq 0$. By graphing many examples (see Figure 2(a)) the students make the connections between the analytic representation (the vertical asymptote $x = -D/C$) and its visual interpretation. Similarly, the analytic horizontal asymptote $y = A/C$ is associated with the flat shape of the graph for $|x|$ large (Figure 2(b)). Both asymptotes (lines) $y = A/C$ and $x = -D/C$ can be added to the graphs for visual support.

Figure 3 shows a typical case for a rational function with numerator degree 3 and denominator degree 1. Notice the graphical end behavior analysis leads naturally to a new concept of *end behavior model* ($y = x^2$ is an end behavior model for $y = (x^3 - 10x^2 + x + 50)/(x-2)$). This is the mathematics concept of *global analysis*. Furthermore, the polynomial division algorithm leads to the definition of a unique *end behavior asymptote*. From the division algorithm, $(x^3 - 10x^2 + x + 50)/(x-2) = x^2 - 8x - 15 + 20/(x-2)$. This purely analytic activity has a wonderful geometric interpretation shown in Figure 3(c). Notice how the quadratic graph most closely "matches" the rational function for all values of x except those near 2!

Here students begin to build their own intuitive understanding of the behavior of different classes of important functions. The rich intuition developed in precalculus aids in the analytic study of functions in calculus.

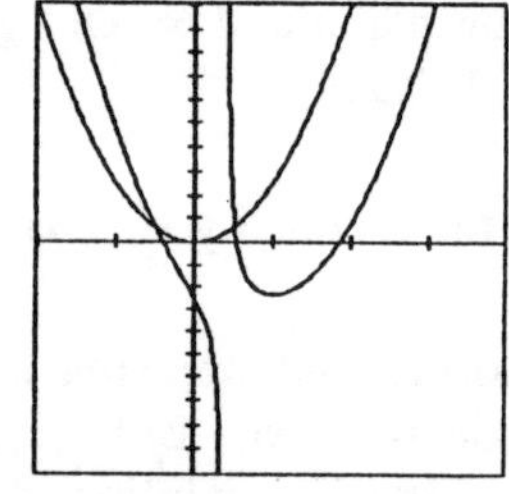

(a) $[-10, 20]$ by $[-100, 100]$

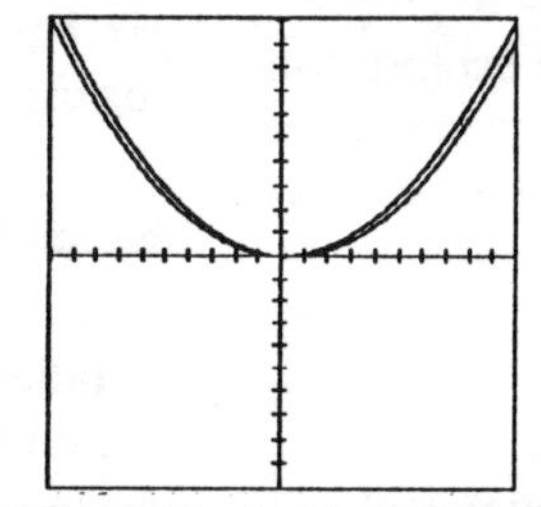

(b) $[-100, 100]$ by $[-10{,}000, 10{,}000]$

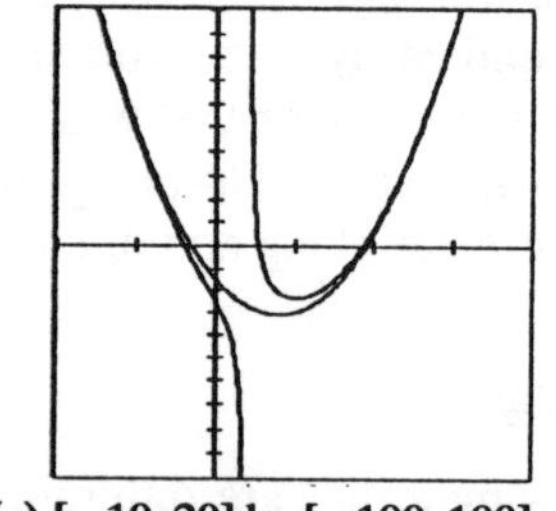

(c) $[-10, 20]$ by $[-100, 100]$

Figure 3: Two views of $y = x^2$ and $y = f(x) = (x^3 - 10x^2 + x + 50)(x-2)$ are shown in (a) and (b), while (c) also shows the graph of the unique quadratic asymptote, $y = x^2 - 8x - 15$, to the graph of $y = f(x)$.

Recent Developments in Secondary School Mathematics, and Their Implications

Zalman Usiskin
University of Chicago

Definitions of terms

In this essay "secondary school mathematics" refers to the mathematics that is taught or should be taught in grades 7–12, which might be broadly conceived of as mathematics beyond arithmetic and the basics of measurement and geometry.

Developments in secondary school mathematics can be considered from a variety of levels. The Second International Mathematics Study [25, pp. 5ff] identified three such levels: the *intended curriculum,* that is, the ideal curriculum as represented by the goals specified in reports, recommendations, and curriculum guides; the *implemented curriculum,* that which is taught; *the achieved curriculum,* that which is learned by students. The personnel involved in these three levels differ: college level mathematicians and mathematics educators and leaders in school mathematics are usually those who set the ideal curriculum; teachers are the main players in the implemented curriculum; students are the determiners of the achieved curriculum. A fourth major role is played by all those who are engaged in the development of products that attempt to translate the ideal into the intended or achieved, including publishers, hardware and software developers, other materials developers, and many engaged in various curriculum projects. These products constitute what we might call the *available curriculum.* A fifth role involves the various aspects of assessment and testing, and has been called the *tested curriculum.*

The phrase "recent development" is a relative term; schools, school systems, and occasionally entire states change at different times in different ways and at different rates. For many schools, changes take place almost exclusively at adoption time, once every five or six or sometimes more years. In times like the present, when major changes have occurred in the technology and other materials available to schools, and major recommendations for reform are contained in the reports of national organizations and of education agencies in many states, the differences between schools that have made changes and those that have not can be dramatic. For instance, some schools today make graphing calculators available to students in all their mathematics courses while others forbid the use of any calculators. Some schools are teaching algebra to a majority of their 8th graders while others do not give any students this opportunity.

In this essay, "recent development" is intended to refer to any development which might lead to students entering calculus with substantially different backgrounds and experiences than they presently have. Since it takes at least a year for a school system to make changes, and six years to traverse secondary school mathematics, a school system would have had to make changes in 1986 in order to have its 1993 high school graduates encounter a different experience throughout secondary school, and thus that system would have to rely on reports and materials written even earlier. This essay considers developments from 1985 on as recent,[1] and is organized by the five types of curriculum identified above.

The ideal curriculum

Thinking at the level of the ideal curriculum in 1985 was greatly influenced by NCTM's *Agenda For Action* [14], a brief document that recommended a broad curriculum at all levels centered around problem solving, and by the CBMS report "The Mathematical Sciences Curriculum: What Is Still Fundamental and What Is Not?" [3], which questioned some of the sacred cows of content in the curriculum, such as triangle trigonometry, and recommended greater attention to discrete mathematics, statistics, and applied mathematics. These and other reports greatly influenced the NCTM *Curriculum and Evaluation Standards for School Mathematics* [15]. This document is no doubt well-known to readers of this essay, but because of its influence (some aspects of which are described below), it is useful to identify some of its major features.

[1] For a summary of earlier developments, see Usiskin [27]. The existence of this earlier essay is another reason why I picked the year 1985 as a starting point.

Here is a list of the sections of the document in its discussion of grades 9–12.

1. Mathematics as Problem Solving
2. Mathematics as Reasoning
3. Mathematics as Communication
4. Mathematical Connections
5. Algebra
6. Functions
7. Geometry from a Synthetic Perspective
8. Geometry from an Analytic Perspective
9. Trigonometry
10. Statistics
11. Probability
12. Discrete Mathematics
13. Conceptual Underpinnings of Calculus
14. Mathematical Structure

The first three of these sections of the *Standards* present a broad view of mathematical thinking. However, there is a notable omission, namely the view of mathematics as procedures. A de-emphasis on complicated paper-and-pencil manipulations is mentioned, but there is no in-depth discussion of the role of technology that can do manipulative algebra, not much said about what students should be able to do mentally, what paper-and-pencil skills they should have, and what they should learn to do with technology.

The fourth section of the *Standards* emphasizes the desire not only to decompartmentalize mathematics courses (despite the compartmentalization in sections 5 through 14) but also to connect mathematics with the other subjects taught in these grades. This goes along with the general flavor of the document towards an integrated curriculum, one in which the teaching of algebra and geometry is not done in full year courses but is spread out over all years. Integration was considered necessary in order to give time for the many topics not now represented in the curriculum but recommended in the later sections. An earlier curriculum that teaches many topics in each year is the New York State Regents' curriculum [20].

The recommendation for technology in the *Standards* is strong: calculators are recommended at all levels, with increasing sophistication for older students. A general call for significant attention to applications of mathematics permeates virtually every section.

A second NCTM document, *Professional Standards for Teaching Mathematics* [16], deals with instructional process and is quite a bit more ambitious and groundbreaking. The teacher is asked to change roles from being an explicator to being a facilitator who runs a student-centered classroom in which discussion about mathematics by and among students is encouraged. A nontrivial outgrowth of this recommendation is that in such a class it is natural that students will spend more time in exploring, discovering, and inventing mathematics, thus spending more time engaged in inductive rather than deductive processes.

The year 1985 was also the first year for the Mathematical Sciences Education Board. The first policy document in which MSEB played a role, *Everybody Counts* [17], brings another theme to the surface: the need to increase the pool of students who take significant amounts of mathematics. In a growing number of places (including Louisiana, Mississippi, and the city of Chicago), all students are required to successfully complete a year of algebra and a year of geometry to graduate from high school.[2]

The influence of MSEB at this level of curriculum cannot be ignored. The concordance of the MAA report *A Call for Change* [8] with the NCTM *Professional Standards* was engineered by MSEB. *Reshaping School Mathematics* [10] gives the philosophy behind the movement for change and *Counting On You* [11] summarizes the various recommendations of organizations at both the precollege and college levels. The current NCTM president reports that the influence of these documents on state and local curriculum guides has been substantial [7].

The available curriculum

Technology

By 1985, inexpensive scientific calculators had already been on the market for a few years, but there was not widespread usage of them in mathematics classrooms. In 1985, Casio introduced a graphing calculator[3] appropriate for use by high school students, followed in 1986 by a similar Sharp calculator.

The Texas Instruments TI-81, a more user-friendly graphing calculator, appeared in 1990. The phenome-

[2]It is an interesting application (or variant) of the principle of compensatory education that these systems, which have some of the weakest students in the nation as judged by college entrance tests, now have the strongest requirements on paper.

[3]Casio and TI call these "graphics calculators," but most users refer to them as "graphing calculators."

nal reception that this calculator has received has been countered by newer Casio and Sharp models, and a newer Casio calculator with a capability between a standard scientific calculator and these graphing models. The capabilities of these calculators are well-known: they enable every real function studied in calculus to be graphed with domains chosen by the user; they enable zeros, maxima and minima, and points of inflection to be estimated using a trace feature; some even enable the user to obtain estimates of derivatives at a point and definite integrals. Newer more powerful models continue to appear, and the trend of putting capabilities into hand-held devices that hitherto have only been available on desktop computers is certain to continue. The advantage for teaching is well-known: students can take these devices home with them, and thus the instructor can give assignments involving technology without worrying about inequity of accessibility to technology.

These calculators do not perform symbolic algebraic manipulations; for that one needs either to go to Hewlett-Packard calculators or to computer software. Because of their complexity and cost, HP calculators are rarely used in pre-college classrooms. The computer software, found more than the calculators but still seldom used, is of two types: software such as *Derive* or *Mathematica* that is appropriate more for the sophisticated user than the neophyte, or weaker software such as the *Mathematics Exploration Toolkit* that is appropriate for student learning. It is likely that within a few years some less expensive and more user-friendly calculators will have symbolic capabilities.

In 1985, one software program, the *Geometric Supposer,* was available for drawing and explorations with geometric figures in the plane. Others have since appeared, including *GeoExplorer,* the *Geometer's Toolkit, Cabri,* the *Geometric SuperSupposer,* and *The Geometer's Sketchpad.* These make it possible for students to explore geometrical figures and properties by generating examples quickly and easily.

This brief summary does not do justice to the variety of computer software currently available for dealing with traditional algebra skills, geometry proofs, statistics, and functions. Furthermore, the sophistication of the available software continues to grow as the power and memory of lower-end computers increases.

Written materials

Written materials have generally lagged behind the available technology. The first widely-available materials for younger students that *required* scientific calculators were those of the University of Chicago School Mathematics Project [26]. The six years of UCSMP materials, designed for grades 7–12, cover the broad spectrum of content recommended in the NCTM *Standards,* with applications throughout, and including substantial amounts of work with statistics and discrete mathematics. The broad spectrum of mathematical topics, but without the applications or technology, is represented in several series of books written for the New York State Regents' syllabus (see, e.g., [2]).

To my knowledge, the first materials at the precalculus level that took into account the available graphing technology were those developed by Waits and Demana, now in its second edition [4]. The last two UCSMP courses require function-graphing technology; the fifth course requires technology for statistics and simulations. These requirements are likewise found in precalculus, statistics, and discrete mathematics materials developed at the North Carolina School for Science and Mathematics [18,19]. The Computer-Intensive Algebra project [5] utilizes *Derive* in its first-year algebra materials.

A more exploratory approach to geometry (emphasizing inductive processes, deemphasizing deduction) is found in a book by Serra [24]. The algebra texts of Benson et al. [1] and the Hawaii Project [23] have increased numbers of real-world applications and a decreased emphasis on manipulative skills.

An integrated problem-solving approach in line with the *Professional Teaching Standards* is found in the materials of the Interactive Mathematics Project (in preparation), one of four high school curriculum-development projects funded by NSF within the past year, joining five middle school projects funded in 1991.[4]

The MAA and NCTM have jointly sponsored materials particularly for the graphing calculator [9]. And there exist many publications—too many to mention—on newer topics in the curriculum, such as discrete mathematics, statistics, fractals, and chaos.

Videotapes

Project Mathematics! [22] has developed a series of videotapes on secondary school mathematics topics for

[4]The other senior high school projects are: the Core-plus Mathematics Project, centered at Western Michigan University; COMAP's Application/Reform in Secondary Education (ARISE); and the Hartford (CT) Alliance Secondary Mathematics Core Curriculum Initiative. The middle school projects are: Michigan State University's Connected Mathematics Project; the University of Wisconsin's Math in Context: A Connected Curriculum for Grades 5–8; the Institute for Research on Learning's Middle-School Mathematics Through Applications Project; EDC's Seeing and Thinking Mathematically; and the University of Montana's Six Through Eight Mathematics (STEM).

use in secondary school classrooms. There are isolated other tapes available, such as those on fractals [21].

The intended curriculum

None of the above developments has any implications for the mathematics education of students unless it reaches them. So it would be useful to know what of the above developments has reached the classroom. Unlike automobile manufacturers, who make public each week the sales of each model of each car, those in the educational materials business closely guard their sales figures. Only in certain states are these figures collated and published. Thus, in general, we do not know how many copies of any textbook or calculator or piece of software are in the schools, and once in the schools we cannot be certain if or how these materials are used. For the most part, we can only judge by those states and districts that do publicize sales, by topics discussed at professional meetings, by hearsay, and by intuition from comments of mathematics teachers and leaders.

Judging from my personal knowledge of the usage of UCSMP materials (the usage is quite high), I estimate that about 30% of the sales of new (not replacement) mathematics texts in the nation for college-bound students (i.e., from algebra upward) over the past two years are from schools making significant changes in their curriculum in their adoption of new texts. This percent seems to be increasing. However, since only about 1/5 of schools adopt textbooks in any given year, and some delay adoption, the total percent of schools who change in a year may be only 4–6%. Thus in the three years since the appearance of the *NCTM Standards,* perhaps about 15% of students in the country have seen their textbook curriculum change dramatically. However, as the above discussion indicates, there are a variety of newer approaches, so it is difficult to make blanket generalizations. Even though many teachers change with their textbooks, some continue to teach their old courses out of mismatched texts.

The use of graphing calculators has dramatically increased in the past two years. I do not have any hard data from which to estimate a percent of classes or students who are affected. If I were pushed to guess, however, I would say that at least a quarter of all new adoptions of precalculus texts in the past two years have included a technology component, so 20%–30% of precalculus classes are now utilizing this technology at some time, with the usage being lower in states that adopt textbooks statewide, and higher in other states.

Concerning software, function-graphing software has been available for the longest time, and the *Geometer's Sketchpad* has had great influence and a large number of devotees, but again I have no data in terms of percents of students who ever use this or other software. The usage of drill-and-practice software is not insignificant, but does not constitute any change in the curriculum. Together I would estimate that no more than 10% of secondary school mathematics classes make use of computers in any way that would cause their students to be different upon leaving school.

I do not wish even to hazard a guess regarding the number of teachers who have changed their teaching style to follow the *Professional Standards for Teaching Mathematics*; it would be difficult to judge how much change in behavior is necessary to qualify. I think it is safe to say that many mathematics teachers are experimenting with cooperative learning, with group activities, with projects, with student writing, and with many of the other techniques that are promoted in these *Standards.* After the 1970s and 1980s in which discovery learning was somewhat out of fashion, we are back to the 1960s, a time when such learning was encouraged.

There is a great range in the attention given to proof-writing and mathematical systems in the various new curricula. The ease with which numerical and geometric examples can be generated by students makes a formal proof less of a convincing argument than it has been (and for many students it has never been very convincing). One of the unsolved questions in mathematics curriculum and instruction is how to take advantage of the increasing power at a student's disposal to generate examples while at the same time emphasizing the mathematical necessity that examples do not suffice as a proof of a general statement.

The tested curriculum

Curriculum (the content), instruction (the delivery system), and testing (the evaluation system) are the three main components of mathematics teaching. The tested curriculum is as complex as the other two, and one could consider an "ideal tested curriculum," an "intended tested curriculum," and so on. Here we group these together.

Students in schools today are tested in a variety of ways, among them: standardized test batteries (either from test publishers or, increasingly, from states) in which the level of mathematics is low; standardized tests designed to cover specific courses; teacher-made tests and quizzes throughout the year of teaching; and the SAT and ACT college entrance tests. It is not an exaggeration that for some classes, 20% of instructional time is devoted to tests of one of these types.

There is as much a desire from some leaders to change the tests as to change the curriculum. Whereas in the past the desire for change in tests was justified by the need for tests to be aligned with curriculum, there is presently in some quarters a conscious purpose to change the tests so that the curriculum will change to be aligned with them, and so that the time occupied by tests can be part of the student's learning experience. Several reports have addressed these issues [15, 12], and it has been announced that a there will be a third volume of the NCTM *Standards* series devoted to assessment standards.

Three kinds of changes are being promoted for tests: (1) changes in the content coverage to represent the wider range of mathematical activities promoted by the standards; (2) changes in the technology allowed, including provisions for calculator or computer use; and (3) changes in the kinds of instruments and modes of assessment to include out-of-school experiences, essays, open-ended questions, projects, and portfolios. Those who promote change generally assume that because calculators and computers are widely used outside of school to perform routine computations and other procedures, assessment will need to concentrate more on higher order processes [6].

The most widely publicized of all changes in testing has been the accommodation of calculators on the various mathematics exams of the College Board. The PSATs already permit four-function, scientific, and graphing calculators; the SATs will allow them beginning in March 1994. Beginning in May 1994, students will have the option to take a calculator-dependent Level II mathematics achievement test. The AP calculus exams will essentially require scientific calculators beginning next month, and will probably require graphing calculators beginning in 1995. These moves can only accelerate the amount of technology utilized in secondary school classrooms. Less publicized is the fact that some standardized test batteries are now available in calculator versions.

Teacher-made tests reflect the variety of practices in use in schools today. Some schools allow calculators on all exams and test at certain times with computers, while others forbid their use. The trend seems rather clearly in the direction of greater use of technology in all sorts of testing.[5]

[5]This is not as revolutionary a development as it might seem if one considers that the paper-and-pencil algorithms themselves constituted a new technology when they were first developed.

The achieved curriculum

There is very little data concerning the effects of the changes in precollege mathematics education that have been recently recommended or implemented. There has not been enough time to allow students to proceed through a curriculum such as those being recommended. For instance, the first students to complete the entire 6-year UCSMP secondary curriculum graduated only last year, and at least half of the books used by these students were preliminary test versions.

Allow me to speculate in this last section about the future. Although some of the impetus for change in the mathematics curriculum has come from low student performance both within the United States and in comparison with comparable students from other countries [13], the only changes recommended for the curriculum likely to increase *mean* performance on traditional tests are those which have the effect of increasing the availability of advanced courses to students. The trends are positive: AP calculus enrollment continues to increase, and recent data indicates that 16–17% of 8th graders were enrolled in algebra in 1992 versus approximately 13% in 1981, thus suggesting the larger pool will continue. But this increase in the pool may result in many students who have lower individual performance than students in the past.

Now let us ask what changes colleges can expect from students who have experienced those schools and classrooms in which significant change has been made. As one might expect, the benefits do not come without cost.

- Students will be more than comfortable with technology; they will expect to be able to use it. They will not see much reality in questions that can easily be answered with technology but for which they are expected to operate without it. The technology will increase their knowledge of graphs of functions, and transformations of such graphs (and perhaps of transformations and matrices in general). They will be computer literate, but they may be less likely to know a programming language.

- Students will be accustomed to seeing applications in their mathematics courses, including curve-fitting and other dealings with data, and they are more likely to be bothered if there are no applications in a course. Thus the increasing tendency for students to acquire college-level mathematics in departments other than mathematics may continue.

- For two reasons, students will be less proficient in algebraic manipulation than they were a generation ago. First, they will not have had courses which emphasize skills and little else. Second, they will not be as tolerant of learning such skills without motivation to learn them. Yet they may be better able to deal with complicated functions and messy numbers.

- Because they will have been instructed with less lecture, students may be less tolerant of lecture courses, or less practiced in taking notes. However, students may be more self-reliant.
- If students have been accustomed to broad-based assessments, they are likely to be bothered by courses in which the entire grade is determined by a single exam, in which they are not allowed to exhibit what they know through oral presentations or reports. They may not be as savvy as current students on multiple-choice tests.

The small amount of data we have from our 6-year UCSMP students can be summarized as follows. Scores on ACTs and SATs seemed about what one would expect from these students. College mathematics placement exams were difficult, and students were often placed in precalculus even though they had two years of such work. When the high school intervened, the students were put into calculus, in which they all are doing A or B work. In short, for these students the college placement tests seem to be out of sync with the prerequisites needed for calculus. Some students opted for finite mathematics rather than calculus, and all felt they knew more than their classmates. The NCTM *Standards* recommend that high school mathematics prepare students for more than calculus at the college level; if current calculus courses do not change, one result of this broader background may be that students will enter the mathematical sciences through courses other than calculus.

References

1. Benson, John, Sara Dodge, Walter Dodge, Charles Hamberg, George Milauskas, and Richard Rukin. *Algebra 1. Algebra 2 with Trigonometry.* Evanston, IL: McDougal, Littell, 1991.
2. Bumby, Douglas and Richard Klutch. *Mathematics: A Topical Approach. Course 1. Course 2. Course 3.* Columbus, OH: Charles Merrill, 1985.
3. Conference Board of the Mathematical Sciences. *The Mathematical Sciences Curriculum K-12: What Is Still Fundamental and What Is Not.* Report to NSB Commission on Precollege Education in Mathematics, Science, and Technology. Washington, DC: National Science Foundation, 1982.
4. Demana, Franklin, Bert K. Waits, and Stanley R. Clemens. *Precalculus Mathematics: A Graphing Approach.* Reading, MA: Addison-Wesley, 1992.
5. Fey, James, and M. Kathleen Heid. *Computer-Intensive Algebra.* College Park, MD and College Station, PA: The University of Maryland and Pennsylvania State University, in progress.
6. Kulm, Gerald, editor. *Assessing Higher Order Thinking in Mathematics.* Washington, DC: American Association for the Advancement of Science, 1990.
7. Lindquist, Mary Montgomery. Shaping the Vision of the Standards. *National Council of Teachers of Mathematics News Bulletin.* Reston, VA: NCTM, May, 1992, p. 3.
8. Mathematical Association of America. *A Call for Change: Recommendations for the Mathematical Preparation of Teachers of Mathematics.* Washington, DC: MAA, 1991.
9. Mathematical Association of America and National Council of Teachers of Mathematics. *Teaching Mathematics with Calculators: A National Workshop. The Graphing Calculator: Building New Models. The Fractions Calculator: Old Things, New Ways.* Washington, MAA, 1992.
10. Mathematical Sciences Education Board. *Reshaping School Mathematics.* Washington, DC: National Academy Press, 1990.
11. Mathematical Sciences Education Board. *Counting On You.* Washington, DC: National Academy Press, 1991.
12. Mathematical Sciences Education Board. *Measuring Up.* Washington, DC: National Academy Press, 1993.
13. McKnight, Curtis, et al. *The Underachieving Curriculum: Assessing U.S. School Mathematics from an International Perspective.* Champaign, IL: Stipes Publishing Co., 1987.
14. National Council of Teachers of Mathematics. *An Agenda for Action.* Reston, VA: NCTM, 1980.
15. National Council of Teachers of Mathematics. *Curriculum and Evaluation Standards for School Mathematics.* Reston, VA: NCTM, 1989.
16. National Council of Teachers of Mathematics. *Professional Standards for Teaching Mathematics.* Reston, VA: NCTM, 1991.

17. National Research Council. *Everybody Counts.* Washington, DC: National Academy Press, 1989.

18. North Carolina School of Science and Mathematics. *New Topics for Secondary School Mathematics. Data Analysis. Geometric Probability. Matrices.* Reston, VA: NCTM, 1988.

19. North Carolina School of Science and Mathematics. *Contemporary Precalculus Through Applications.* Providence, RI: Janson Publications, 1991.

20. Paul, Fredric, and Lynn Richbart. New York State's New Three Year Sequence for High School Mathematics. In *The Secondary School Mathematics Curriculum,* edited by Christian R. Hirsch and Marilyn Zweng, the 1985 Yearbook of the National Council of Teachers of Mathematics. Reston, VA: NCTM, 1985.

21. Peitgen, H.-O., H. Jurgens, D. Saupe, and C. Zahlten. *Fractals: An Animated Discussion.* (Videotape) San Francisco: W.H. Freeman, 1990.

22. Project Mathematics!. *The Pythagorean Theorem.* (Videotape) Pasadena, CA: California Institute of Technology, 1988.

23. Rachlin, Sidney, Annette N. Matsumoto, and Li Ann T. Wada, *Algebra I: A Process Approach.* Honolulu: University of Hawaii Curriculum Research and Development Group, 1992.

24. Serra, Michael. *Geometry.* San Francisco: Key Curriculum Press, 1992.

25. Travers, Kenneth J. and Ian Westbury, editors. *The IEA Study of Mathematics I: Analysis of Mathematics Curricula.* Oxford: Pergamon Press, 1989.

26. University of Chicago School Mathematics Project. *Transition Mathematics. Algebra. Geometry. Advanced Algebra. Functions, Statistics, and Trigonometry. Precalculus and Discrete Mathematics.* Glenview, IL: Scott Foresman, 1990, 1991, 1992.

27. Usiskin, Zalman. We Need Another Revolution in Secondary School Mathematics. In *The Secondary School Mathematics Curriculum,* edited by Christian R. Hirsch and Marilyn Zweng, the 1985 Yearbook of the National Council of Teachers of Mathematics. Reston, VA: NCTM, 1985.

Precalculus and Calculus Reform at Community Colleges

Stephen B. Rodi
Austin Community College

Sheldon P. Gordon
Suffolk Community College

In this article we will lay out a community college perspective on curriculum reform in precalculus and calculus. The assignment is daunting because of the number and variety of community colleges in the United States and because of the staggering number of community college students who are in precalculus and calculus courses.

Demographics

The American Association of Community Colleges (AACC) counts over 1200 two-year institutions in the United States. The 1990 Digest of Educational Statistics confirms this number and reports that these schools enroll over 6,000,000 students, four times as many as in 1966. This enrollment accounts for over 30% of the full-time equivalent enrollment in colleges and universities across the country and a much higher percentage of "head" count.

About 65% of two-year college students attend part-time. In addition, data from a Department of Education report supports the following conclusion: "Continuing a trend of the 1980's, minority students were more likely than white students to attend two-year colleges. While 38 per cent of white students were enrolled in such colleges in 1991, the comparable figures were 55 per cent for Hispanics and American Indians, 43 per cent for blacks, and 40 per cent for Asians." [4, p. A30]

More specifically, the most recent mathematics survey (1991) conducted by the Conference Board of the Mathematical Sciences (CBMS) estimates that 1,400,000 community college students are enrolled in mathematics, also a fourfold increase over the comparable 1966 numbers. Of those, 245,000—about 18% of the two-year college mathematics enrollment—are in precalculus, college algebra, or trigonometry courses. The overall precalculus enrollment has grown 260% since 1966, 30% since 1985.

Mainstream calculus courses at community colleges enroll about 94,000 students with other calculus courses enrolling an additional 34,000 students. In some areas, the significance of community college calculus enrollment is dramatically evident. In the state of Washington, for example, one out of every two calculus students takes the course at a community college.

Community Colleges and the Pipeline

These numbers confirm that community colleges are the obvious reservoir from which to draw talent to fill the national scientific pipeline, especially at a time when demographic studies suggest the next generation of mathematicians, physicists, chemists, engineers, computer scientists, and skilled technicians in the United States must come in large numbers from currently underrepresented groups. The National Science Foundation has already recognized this in projects such as its Alliances for Minority Participation (AMP). A major goal of these alliances is to increase the rate at which minority students in science and mathematics transfer from two to four-year schools . One tool to achieve this aim is the summer bridge program. The importance of this transfer rate is discussed in *The Chronicle* article [4] mentioned above.

The Clinton administration already has targeted community colleges in its education reform plans, especially as these plans touch on technology training. Even prior to the new administration, Congress had designated National Science Foundation funds specifically for community college "two plus two" programs in mathematics, science, and technology to create a transition for these students from secondary school to collegiate technically oriented programs.

Our Audience

Many of those involved in mathematics reform projects do not know much about the structure and culture of community colleges. Consequently, these individuals form an important part of the audience we address in this paper.

We believe it is far more common for community college mathematics faculty to understand the world of their four-year colleagues than vice versa for the obvious reason that most community college faculty have spent time in universities earning advanced degrees. (Do we hear an argument here for university graduate students in mathematics to serve internships in community college mathematics programs or for four-year college faculty to have visiting professorships at two-year schools?) In light of this one-way transfer of experience, some of what we do in the rest of this paper simply is an attempt to tell our four-year college readers and others about community colleges.

But our hope is that this paper will be useful to a number of other audiences as well. Community college faculty can use it to stimulate their own thinking and to provide a basis for discussion with their department chairs, deans, and vice-presidents. Community college administrators will find here a catalogue of the issues that should concern both their governing boards and their staffs. Those who write consortia proposals but are unfamiliar with life at the community college can use it to gain insight into relevant issues. The National Science Foundation and other funders might find some ideas in the paper that affect the structure and focus of their programs.

Governance and Mission

Community colleges can be part of tightly organized state-wide systems such as in Virginia, or they can be in states such as California and Florida where the system, though more loosely organized, is still highly structured and funnels large numbers of students to four-year campuses. (California alone has over 100 community colleges.) These colleges might be part of multi-campus local area mega-districts such as those in Dallas or Miami or Maricopa County in the Phoenix area (which enrolls over 100,000 students). They might be like those in Kentucky, Ohio, or Wisconsin which are formally part of the same governance structure as the four-year colleges and universities. On the other hand, many community colleges are like those in Texas, governed almost exclusively by local community college boards which are comparable in structure to local school boards and which have almost complete local autonomy to make decisions for the college. Funding for community colleges usually is proportioned by thirds among student tuition, local taxes, and state legislature appropriations, though there are many variations of this pattern.

It is important to be aware of the governing structure of any particular community college since that can have an important impact on its attitude about and receptivity to curriculum change and on the ease and facility with which the faculty and administration can implement change. On the one hand, local control frequently is an asset since it makes the chain from idea to funding relatively short and allows for more on-site flexibility. On the other hand, one must sometimes deal with the narrower viewpoints and agendas that can be part of local politics. In large districts, curriculum decisions can be cumbersome if they require achieving a consensus among many separate campuses; but such districts may have the resources to make significant impact over a wide geographical area affecting many students.

Community colleges generally serve far broader needs than most four-year institutions. Most are known as comprehensive community colleges, implying that within a single college structure one will find university transfer curricula (equivalent to freshman/sophomore courses or programs at four-year schools); work-place oriented vocational and technical programs which lead to certificates and associate degrees and whose purpose is to directly prepare students for jobs; centers for the assistance and promotion of local business and industry; GED programs for high school equivalence diplomas; and extensive developmental studies programs which offer a wide range of precollege courses from study-skills to basic algebra. This comprehensive structure means that community college administrators have varied and sometimes antagonistic constituencies, many of them nonacademic by four-year college standards.

The complex mission of community colleges can have a complicating effect on the precalculus curriculum in particular. The precalculus program frequently must serve two different student populations who take the courses with different intent. Some students in college algebra and trigonometry are aiming at calculus and intend (for example) to transfer to engineering degree programs at four-year schools. Others are in terminal associate degree programs such as engineering technology, electronics, air conditioning, drafting, or surveying. It is not always easy to design a single sequence of courses to serve both student populations; and smaller community colleges, like smaller liberal arts colleges, may not have the enrollment to justify two separate sequences which deal with precalculus material.

Venues for Reform

One should not conclude from the complexity of community college governance and mission that two-year colleges are not good venues for precalculus and calculus experimentation and reform. Quite the contrary, community college mathematics departments have many characteristics which make them excellent candidates for experimental efforts.

First and foremost, teaching is what community colleges are all about. There are very few competing obligations. The reward and promotion system based on research in place at most universities is inappropriate for the community college and rarely is part of the promotion process. Faculty rise to the top of their professions at community colleges by being outstanding teachers and are rewarded on that basis. Community colleges are naturally attuned to the extended definition of scholarship offered by Ernest Boyer of the Carnegie Foundation, which includes as scholarly work the synthesis and presentation of existing knowledge.

Because their professional focus usually is exclusively on teaching, community college faculty frequently are more adaptive, more open to experimentation, and more willing to take risks in instructional matters than their four-year colleagues. Such response and mobility is part of the overall community college mission and one of the characteristics by which a community college defines itself to its community. At the same time, it is creative activity centered on the classroom that provides two-year college faculty with the professional and personal growth that research provides in the university environment.

There is a long history at community colleges of concern about, and experimentation with, student learning styles and of attempts to adjust instruction to reflect how students learn. Such ideas usually find a receptive audience with community college administrators. Community colleges have a tradition of worrying about the student holistically, using evaluation of the student's background, outside obligations (e.g., work, family), and personal goals to structure comprehensive recommendations about the student's plan of study.

The divisional structure in community colleges—wherein a number of related disciplines such as mathematics, the sciences, computer science, and engineering are grouped in the same organizational unit—can facilitate technological and applications-oriented approaches to instruction. For instance, several two-year colleges currently are experimenting with year-long integrated courses in calculus and physics which are team-taught by faculty from both disciplines. In addition to this internal organizational support, students and faculty at community colleges usually have a certain pragmatism, which readily supports an applications viewpoint.

It is not unusual to find more computers, computer labs, computer-assisted instruction, and computer-comfortable faculty at community colleges than at neighboring four-year schools. This is a strength that can be built on in both precalculus and calculus reform. What may be lacking in community colleges is access to, and understanding of, more sophisticated mathematics packages such as *Maple* or *Mathematica* or the sophisticated computer workstations now available at many universities. But lack of access to software tools or training in their use frequently is an easier problem to solve than the purchase of hardware, particularly at two-year colleges where desire to use existing computer systems seems to come naturally to faculty.

Other features of the community college instructional program support experimentation. Class sizes are small, rarely over 35 students. CBMS reports that the average precalculus class at a community college has 27 students. Tutoring labs are common, estimated by CBMS to exist at 86% of all two-year colleges. Instructors are directly in charge of and in contact with all aspects of their classes, including labs, so that part of the instruction is not turned over to a third party such as a teaching assistant.

There is a mind-boggling wealth of experience about precalculus, in particular, in almost every full-time community college faculty member. Someone told the story about a recently retired university faculty member, a distinguished leader in the current calculus reform efforts, who had taught calculus 22 times in his forty-year career. For senior faculty at a university, this actually is an impressive base of experience. However, at a community college it would not be unusual for a faculty member to teach college algebra 22 times in less than five years!

On a related note, teaching a course too often leads to boredom, routine, monotony, and even burn-out. All these are reasons why community college faculty are ready and even eager to launch experimental projects on their own. They also are reasons why reforms efforts, which in calculus already have been shown to revitalize faculty, are particularly important in community colleges.

Finally, students at two-year colleges tend to be weaker, at least in terms of their algebraic skills, than students at four-year schools. Consequently, success rates in precalculus level courses are often considerably worse at two-year schools. Therefore, faculty are more anxious and motivated to find better ways to prepare their students for calculus and to encourage students to move to higher level mathematics courses. This certainly creates fertile ground for curricular reform.

Evidence of two-year receptivity to curriculum reform continues to grow. Since the Allerton Conference for which this article was originally written, AMATYC has been conducting a major project to develop a set of curriculum standards for all courses below the level of calculus. The spirit of the recommendations, which call for dramatically different content and focus in these courses, is consistent with both the NCTM *Standards* and the calculus reform movement. The recommendations call for an infusion of new mathematical ideas into the curriculum based on real-world data and graphical analysis in all courses, an increased emphasis on meaningful applications, and more focus on understanding fundamental mathematical concepts instead of a predominant focus on developing routine manipulative skills. The final report is scheduled for publication by the end of 1994.

Special Problems At Community Colleges

But community colleges have their own special problems which can interfere with reform and need to be faced before reform can proceed successfully. Some of these are serious, indeed. We lay them out here, not necessarily in order of importance, since their relative weight might vary from institution to institution.

Many community college curricula in precalculus and calculus are transfer-driven. Enrollments remain high when students are certain that their courses will transfer directly to nearby senior institutions. When a community college is linked geographically almost exclusively to one university, its curriculum will clone that of the university. In other settings where a community college sends students to many different senior colleges, the faculty has the complicated task of designing a course sequence that will be accepted by all receiving schools.

The reality of transferability can make community colleges cautious about launching experimental projects on their own. Sometimes they are mandated by state guidelines to follow the same course structure as senior colleges. At a minimum, they look over their shoulders fearing that too significant a divergence from big brother's patterns will shut down their own student supply or at least not serve their transfer students well. Often community college faculty feel that it is not possible for them to make any kind of impact outside of their own institutions, particularly on the national scale, but even on a regional one. They may doubt that there is any realistic chance of attracting funding to allow them to develop their ideas for a wider audience.

Important advantages of the regional and state-wide coalitions and consortia favored by the National Science Foundation in its calculus dissemination efforts are (1) they directly address questions of transferability and common curricula, and (2) they provide mechanisms and models for community college faculty to use in overcoming their timidity. In fact, the transferability issue disappears when pairs or groups of two and four-year schools work together simultaneously to change curricula. Such pari passu progress should continue to be a high priority in NSF funding of reform projects.

We also note that the academic world these days more frequently is being turned upside down (to paraphrase the title of the song the British army played during their Yorktown surrender to George Washington) with regards to who is following whose reform lead. More and more one sees community colleges taking the initiative in reform, sometimes with nearby four-year schools then falling in step. Some examples:

- Suffolk Community College has served as a node to attract a variety of four-year colleges and universities into reforming their calculus offerings and adopting the Harvard curriculum.

- Seattle Central Community College and Evergreen State College were partners in a model state-wide curriculum experimentation and dissemination project for Washington State, funded by the National Science Foundation.

- Nassau Community College was among the very first schools in the country (two or four-year) to turn an entire calculus curriculum, involving more than a dozen sections per semester, over to one of the reform curricula (Harvard). As a result, their faculty have been lead instructors at training workshops around the nation, including workshops for teaching assistants at schools like the University of Michigan.

- The Borough of Manhattan Community College has developed a cooperative learning environment patterned on the Uri Treisman model for minority students taking calculus, which now serves as a model for the entire CUNY system, both two and four-year institutions.

- Austin Community College sponsored this spring (1993) the first workshop in central Texas (and perhaps in the state) on the Harvard calculus materials.

- Finally, Suffolk Community College has taken the lead in organizing a consortium involving primarily four-year colleges and universities to reform precalculus offerings. (See an accompanying article in this volume for details.)

Other Issues

There is a list of other issues specific to community colleges that reform advocates (and proposal writers) need to consider. We will comment on these in turn below.

Community college faculty have very heavy teaching loads. The norm is fifteen hours weekly in the classroom, but many faculty have considerably heavier teaching loads. This almost always involves three course preparations and not uncommonly a fourth. Faculty involvement in other aspects of the mathematics program (assessment, tutoring, monitoring part-time faculty) is significant. These heavy instructional loads reduce both the time and energy available for curriculum development or for participating in training workshops. In any reform effort, time (i.e., release time) may be the most significant need of regular full-time community college faculty.

A related factor is the availability of funds in community colleges for in-service training, attendance at conferences and workshops, and sabbaticals. Faculty often complain that their administrations do not understand the need for them to keep up-to-date in their rapidly changing fields. In this respect, the culture within community colleges frequently is more like that of school districts (a little money for a conference every two or three years; emphasis on local staff development activities presented by in-house personnel) rather than the more regular support for travel and other academic events found in senior institutions. Hence, any planned reform should pay special attention to providing funds for, or to encouraging two-year colleges themselves to fund, a broad range of development experiences for two-year faculty.

Training is particularly important for several reasons. Typically, two-year college faculty have been out of school for many years and are in an environment which is not on the cutting edge of mathematics. Many complain that they have lost the mathematical skills and knowledge they once had. Yet, much of the reform activities involve new technologies and new (or at least not well-known) topics in mathematics. Frequently, there is no one available at a community college knowledgeable about either the technology or the mathematics. For instance, several of the calculus reform projects make heavy use of the slope or direction field associated with a differential equation, a topic many community college mathematics faculty have never seen. When no one is around who is comfortable with such ideas, faculty quickly become intimidated and shy away from reform approaches.

Even if the desired reform activities involve nothing more than adopting very simple technology such as a graphing calculator, training is essential. It is not adequate, either at two or four-year institutions, for a faculty member to be handed a calculator and told to use it in the course. The presence of the calculator fundamentally changes the nature of the mathematics being taught. Most instructors cannot be expected to understand the full implications of these changes on their own. Too often we hear stories of faculty members who graft minimal calculator applications onto their courses, ending up with a hodgepodge which satisfies no one, least of all themselves. We cannot emphasize too strongly the need for workshops and training to prevent such occurrences.

Part-time Faculty and Learning Centers

Part-time adjunct faculty make up a significant cohort within the community college mathematics faculty. The 1991 CBMS survey estimated that 48% of all mathematics class sections in community colleges are taught by part-time faculty. CBMS reports that this is a dramatically growing number; it was only 28% in 1985. More significantly, as we will comment on below, this 48% figure is misleading for precalculus courses where part-time faculty teach proportionately more of the class sections. At some schools, such as Austin Community College, the percent of total class sections taught by adjunct faculty has ranged as high as 70%, requiring an unusually large devotion of resources and a wide range of time-consuming activities to keep these faculty oriented and integrated into the mathematics program. These are resources of time and energy that are then not available for curriculum creativity. In California, the aptly named 'freeway flyers' are part-time faculty who actually teach at two or three different community colleges.

There is a strong tendency for part-time faculty to teach as they were taught, even more so than with regular faculty, since part-timers are not in day-to-day contact with pedagogical and curriculum revision at the college where they teach. And they have less time and financial motivation to try something new. A salary of $1200 per three-credit semester course (1/3 less in many places) is a very small carrot with which to entice a part-time worker to use new pedagogy that will require substantial additional preparation time. Indeed, the department chair at a community college who pushes too hard on these matters runs the risk of losing a much needed teacher. Hence, schools which are serious about reform must provide incentives for their part-time faculty to undertake the new and time-consuming personal training and course preparations which these assignments require.

CBMS estimates that 83% of all mainstream calculus class sections at many community colleges are taught by regular full-time faculty. For full-time faculty mainstream calculus is a plum teaching assignment, the equivalent of teaching graduate courses at the university. CBMS also reports that almost twice the proportion (30%) of precalculus classes are taught by part-time faculty. Because of the large enrollment at the precalculus level, this means that about five times as many precalculus classes as calculus classes are taught by part-time faculty at community colleges.

As mathematics reform spreads to the precalculus level, more and more reformed classes will be taught by part-time faculty. The problems associated with their use will be significantly exacerbated. It is imperative that ways be found to fund the training of part-time faculty to incorporate them into reform planning and implementation.

Another issue for community colleges is learning center staff and student tutors. We have already noted that about 86% of all two-year colleges have such centers or tutoring labs. Like part-time faculty, their employees must be brought into mathematics reform in such a way that the support they offer students reinforces rather than interferes with new ways of thinking, teaching, and approaching problems. Here is a relevant vignette. In the Harvard calculus materials, emphasis is placed on the ability to differentiate a function graphically given only the graph of the original function. When students at Suffolk Community College went to their Math Learning Center for help with such questions, the students working there (including some graduate students from a local university) did not know where to start. They demanded a formula for the given function which they could first differentiate and from which they could then sketch the derivative.

To avoid undermining efforts by the regular classroom instructor to emphasize visual and graphical approaches, teaching assistants at universities likely will need the same kind of training as community college tutors. However, TAs at a university are more likely to have contact during much of their day with their department's curriculum reform efforts than are community college tutors who are hired on an hourly basis and frequently are not as mathematically skilled or current. This suggests efforts to train auxiliary mathematics staff in community colleges will be more pressing than similar efforts in universities.

The separation within some community colleges of developmental mathematics from the regular mathematics department can be an important impediment to reform. In such arrangements, beginning and intermediate algebra courses are placed in a self-contained department, often chaired by a reading or writing instructor. Any serious precalculus reform effort eventually will require modification of these early algebra courses. The mathematics faculty in such developmental departments frequently have terminal bachelor's degrees and are mathematically less well trained than faculty in the mathematics departments. Hence, they bring a narrower mathematics background to understanding the intent and techniques of reform approaches. While many are active in professional mathematics organizations, a significant number are members only of associations of developmental educators such as the National Association for Developmental Education (NADE). Hence, their professional associations will give them less exposure to calculus and precalculus curriculum reform as it has developed up to this point. Style and philosophy of instruction also are issues, since some developmental programs (whether they admit it or not) are strongly influenced by the Back-to-Basics movement, teaching for life survival, and rote-centered skills training.

The Student Body

Finally, the community college student body gives rise to some special issues. These students frequently live on the edge financially and find it hard to invest in expensive graphing calculators and other learning tools. Sensitive to this, community college faculty are more reluctant to require such purchases by students, especially when it is less clear initially if the student will perdure into (say) a transfer engineering program and reap full benefit from the investment. Creative solutions need to be found such as calculator lending programs by media centers or libraries, or a buying and selling program for used calculators imitating what students now do with used books. One innovative approach to this has the college purchase graphing calculators and lease them to students for a semester at a nominal fee with an option to buy the machine outright at the end of the term.

The commuter status of almost every community college student also is an issue. Not only do community college students almost never live on campus, but the vast majority are only part-time students. It also is more common than at four-year colleges for them to work and be married and have families. It is not uncommon for them to work forty hours or more a week and carry full academic loads, no matter how strongly a college advises against such an arrangement. All these factors fragment their schedules and reduce their time on campus, making it harder to incorporate them into the small group and collaborative structures which characterize so much of recent calculus reform and promise to be an even more important aspect of precalculus curriculum development.

Conclusion

We promised at the beginning of this paper to try to make it useful to a variety of audiences. Here is a list of specific concerns that apply to community colleges to which these various readers can look for "instant" reference. Most of the items on the list simply paraphrase ideas discussed above. Many, of course, also are relevant at four-year institutions. To provide a focus, we address the list to community college administrators.

If a community college administration is serious about precalculus and calculus reform, then from a combination of its own resources and outside support it must:

1. provide time and opportunity to retrain faculty in the philosophy and content of new curricula;
2. provide time and opportunity for faculty to attend workshops and courses away from the college about current mathematics curriculum developments;
3. provide faculty with time to develop new curricula;
4. provide faculty and instructional programs with appropriate technology;
5. provide time and opportunity for faculty to become familiar with technology and related software;
6. provide time to develop new test materials which are consistent with and reinforce new instructional approaches and philosophies;
7. provide for the development of appropriate placement instruments which are consistent with the intent of new curricula;
8. provide adequate laboratory facilities, both computer labs and specially designed spaces that accommodate interactive, group-based cooperative learning and group computer use;
9. provide time for departmental dialogue as reform programs are being developed and implemented;
10. provide small class sizes, if that does not already exist;
11. provide incentives for part-time faculty to join reform efforts and training to support their involvement;
12. provide time and opportunity for training tutors and math lab/learning center personnel in new approaches to course content;
13. provide contacts with near-by universities to which community college students might transfer;
14. provide e-mail and connections to Internet to facilitate contact with the rest of the mathematics community and share development ideas;
15. provide opportunity to interact with client disciplines, with the high schools who are the source of their students, and with the local businesses and industry that hire their graduates.

Bibliography

For statistical information about two-year colleges, consult references [1], [2], [3], [5], and [6].

1. Albers, Donald J., Don O. Loftsgaarden, Donald C. Rung, Ann E. Watkins, *Statistical Abstract of Undergraduate Programs in the Mathematical Science and Computer Science in the United States: 1990–91 CBMS Survey,* MAA Notes Number 23, The Mathematical Association of America, 1992.
2. *Community and Junior College Directory,* American Association of Community Colleges, One Dupont Circle N.W., Washington, D.C. 20036.
3. *Digest of Educational Statistics,* National Center for Education Statistics, U.S. Department of Education, Washington, D.C.
4. Evamge;auf, Jean, "Number of Minority Students in Colleges Rose by 9% From 1990 to 1991, U. S. Reports," *Chronicle of Higher Education,* 1/20/93.
5. "Trends in Enrollment in Higher Education by Race/Ethnic Category: Fall 1982 through Fall 1991," U.S. Department of Education, Education Information Branch, Room 300, 555 New Jersey Avenue N.W., Washington, D.C., 20208-5641 (phone 1-800-424-1616).

6. "The Eleventh Annual Status Report on Minorities in Higher Education," American Council on Education (ACE), Publication Department, One Dupont Circle, Washington, D.C. 20036.

Calculus and Precalculus Reform at Minority Institutions

James H. Fife
Lincoln University

Background

The desire to change the way we teach calculus is deep and widespread throughout the mathematical community. The failure rate in calculus courses has been high, and because calculus is a requirement for science and engineering courses as well as advanced mathematics courses, it has in effect become a barrier preventing many students from pursuing careers in science, engineering, and mathematics. For minority institutions these problems take on additional significance. African-Americans and Hispanic-Americans are underrepresented in science, engineering, and mathematics professions. Although African-Americans constitute 10% of the U.S. population, they constitute only 2.6% of the American scientists, engineers, and mathematicians. It is widely recognized that the United States lags behind other industrial nations in science and mathematics education in the schools, and it has been predicted that the U.S. will face a shortage of 560,000 scientists, engineers, and mathematicians by the years 2010. According to an article in SIAM News [3], "The U.S. faces a potentially crippling shortage of scientists and engineers in the very near future .... Most of the recommended solutions include the need to turn to some traditionally neglected groups—[including] members of minority groups—to fill out the future ranks of American scientists, engineers, and mathematicians."

For this reason, the nation's historically Black colleges and universities, and other institutions with large minority enrollments, must have an important role in the current calculus and precalculus reform movement.

With the support of the National Science Foundation, several new models for teaching calculus have been developed, and many institutions are experimenting with pilot projects that use the new curricula. Meanwhile, it has become clear to many of the reformers that, at least in institutions with large populations of underprepared students, any successful reform must begin with precalculus. This view was endorsed by the NSF when it expanded its calculus initiative to include the "Bridge to Calculus." In this paper we will examine what some of the nation's historically Black colleges and universities (HBCUs) are doing in calculus and precalculus reform, and what difficulties they are facing. We will also look at some projects at institutions with large minority enrollments.

Reform projects at HBCUs

Several of the projects that are known nationally have been adopted at HBCUs, most notably the Harvard project [6] and the Purdue project using ISETL [4]. For example, Cheney University in Pennsylvania and Southern University in Louisiana are currently using the Harvard materials in some of their sections of calculus. Some faculty at Howard University are incorporating changes in calculus instruction, based in part on the Harvard model, and in part on the Professional Development Program developed by Uri Treisman for minority students in mathematics [7].

Other Howard faculty and some faculty at Delaware State College are currently testing the Purdue program. While there have been some logistical problems at Howard, the instructors there and at Delaware State are enthusiastic about the program, and the student reactions have been positive. According to Richard Bayne, Howard will continue for a time to teach some sections of calculus using the Purdue approach, while others use the Harvard materials.

Bowie State University in Maryland received a Minority Science Improvement Program (MSIP) grant to develop materials to use in calculus. According to Rebecca Lee, the Principal Investigator for the MSIP grant, the existing materials were not appropriate for Bowie students for several reasons. Bowie students are typically commuting students with full-time jobs who spend little time on campus; for such students, the Treisman model with its intensive and lengthy workshops was not possible. Bowie State does not have a dedicated computer lab for calculus instruction; hence major projects requiring long sessions in the computer lab were not feasible. Also many Bowie students have weak precalculus backgrounds. The materials that were developed at Bowie State include small projects that allow the students to explore ideas presented in class and to discover some mathematics for themselves. The projects are suitable for use with *Microcalc,* a software package more user-friendly that *Derive* or *Mathematica*; the projects take care to reinforce the students' precalculus skills while leading them to explore concepts in calculus.

Spelman College in Atlanta received an instrumentation grant from the NSF to revamp a classroom with 15 networked PCs and an overhead projection system. Currently one section of calculus is taught in the room, using Maple. A traditional text is used, with exercises being adapted for use with the computers. There are no pilot sections of computer-based instruction running throughout the calculus sequence; instead, the individual instructors each semester decide if they want to use the computer or not. Because of this, a student in computer-based Calculus I may take a traditional Calculus II, in a section with students who took traditional Calculus I, or a student in a computer-based Calculus II may take a traditional Calculus III. Therefore it is necessary in the computer-based sections to cover all the traditional topics. Thus the radical changes offered by the Harvard curriculum and others are not possible. Similarly, precalculus at Spelman is a traditional course with drill assignments to be done in a math lab equipped with True Basic software.

Our institution, Lincoln University in Pennsylvania, is initially concentrating on reforming the precalculus courses. Since many of our calculus students take at least one semester of precalculus before taking calculus, and since we have been dissatisfied with our precalculus se- [illegible] we thought it appropriate to begin by reforming [illegible] ourses. There is growing concern throughout the mathematical community regarding the success of the precalculus courses taught in the traditional manner. For instance, Treisman [7] has observed that during one year, out of the 422 students enrolled in a precalculus course, only one student went on to receive a grade of B− or higher in Calculus II. Similarly, Fusaro and Gordon [5] have pointed out that each year, approximately 600,000 college students take a precalculus course; yet only about fifteen percent of them ever go on to **start** calculus. From this, it is clear that the current traditional precalculus courses do not meet the needs of underprepared students, and that precalculus has become the filter in the minority mathematics education pipeline, screening out minority students from entering mathematics or science-based careers even before they consider taking calculus. Thus, it seems clear that reform is needed in the teaching of precalculus, along the lines that have been proposed and implemented for the teaching of calculus.

It is not hard to see why the traditional precalculus and college algebra courses are highly inappropriate for the level of students who usually enroll in them. Some of the reasons are the following:

- Traditional precalculus courses merely attempt to reteach algebra and trigonometry in much the same way these subjects were taught in high school. Students never see any new mathematics, and the old mathematics they see is little more that a repeat of what did not work in high school. As a result, those students who approach the course already afraid of the subject leave with their math anxiety reinforced and whatever desire they had to take more mathematics eliminated. Other students are misled into believing they already know everything in the course, and hence do not take the course seriously. Students become bored at being required to take high school mathematics, or angry at being labelled "underprepared" or "remedial."

- Students are led to believe that algebra is merely a collection of arbitrary rules and procedures to be memorized, like the rules of chess or bridge.

- So-called word problems are highly artificial and fail to stimulate any interest in the students. Textbooks divide word problems into categories and present a recipe for each type. Students never understand that applying mathematics means building a mathematical model for a real-life situation and then using mathematics to analyze the model. It is not surprising that when these students arrive in calculus, they demand to learn recipes rather than principles.

- The notion of function, which is critical to calculus, is presented in precalculus as yet another algebraic topic, usually at the end of the semester. Functions must be the central concept in precalculus mathematics.

In response to these concerns, Lincoln is attempting to revise its precalculus sequence, making the notion of function the central idea, drawing examples from real-world situations, and teaching algebra in the context of solving real-world problems.

Reform projects at institutions with large minority enrollments

Institutions with large minority populations are also experimenting with new methods of teaching calculus and precalculus. One of the most distinctive approaches is being tried at the Community College of Philadelphia. The approach there is to consider only functions that have Taylor approximations, and to use these approximations to define their derivatives. This avoids the need to introduce the notion of limit at the very beginning of the course, and simplifies the discussion of the product rule, the chain rule, and so on. According to Alain Schremmer, such an approach is more effective for underprepared students, students whose English skills are not good, and older students who are returning to school after a period of time.

New Mexico State University, with a large Hispanic population, has tested giving large-scale projects to calculus students [2]. In some sections, students are given a multi-week assignment consisting of one difficult problem. Students work in groups and submit a written report. In other sections, students work on shorter assignments, lasting about a week, that introduce core material. These students also work in groups and submit written themes. According to Douglas Kurtz, the written work done in groups is particularly effective for students with poor English skills. New Mexico State hopes to use these projects next year in its precalculus sections.

Difficulties with reform

Of course, calculus and precalculus reform has not been without difficulties, and the situation is no different at the nations' HBCUs. While the problems at HBCUs are certainly not unique to these institutions, they are perhaps more serious there. The more significant problem is the general lack of resources, both institutional resources and student resources. Many HBCUs lack the physical resources that much of the reform movement requires. Some have no dedicated computer lab for calculus and precalculus instruction, their equipment is inadequate, and they do not have the funds to purchase computers or calculators. Many minority students are struggling to pay for college and cannot afford a $100 graphics calculator.

The faculty at most HBCUs have heavy teaching loads, making it difficult to spend the time required to become familiar with the new approaches to teaching calculus. It is also difficult to schedule the extra class time that some of the new curricula require. At many HBCUs there are no student graders, and classes are large, so faculty are unable to require the large assignments that the new curricula recommend. Support for faculty travel is frequently insignificant or nonexistent, making it difficult for faculty at these institutions to attend national meetings and other conferences.

We close with two concerns that have been raised by others. As David Smith observes in his paper in this volume, calculus reform means more than the addition of technology to a standard course; it means changing the way we teach mathematics. It means teaching calculus more as a collection of ideas than as a collection of recipes. But students have come to expect recipes in mathematics courses and can become angry if the instructor attempts to break the pattern of teaching recipes. Smith suggests that this may be a problem mostly with white male students, but at least some instructors at HBCU's have found student resistance to change more deep-seated there than at majority institutions. The concern is that, in the end, the goals of the reform movement will be forgotten, and that the use of technology will substitute one type of recipe for another, and will replace the old cookbook approach with high-tech cookbooks.

The other concern was raised by Sandra Bayne in a recent article in UME Trends [1]. In describing a planning conference in Washington for a consortium of HBCUs, Bayne remarks that "the principal authors of the new calculus curricula are traditional majority culture academics from schools with a reputation for exclusivity.... The attendees [at the conference], for the most part, were faculty from prestigious historic Black colleges and colleges with open enrollment. Notable by their absence were African-American scholars put off by the term 'underprepared' students and faculty from prestigious white colleges in the region.... [This could create] the perception of benign paternalism."

While these issues are not the focus of this paper, they are raised here because they are significant concerns which need to be addressed in any future discussion of calculus and precalculus reform. They also underscore the point that the nation's historically Black colleges and universities and other institutions with large minority enrollments must have an important role in the calculus and precalculus reform movement.

References

1. Bayne, Sandra Carmichael, "Teaching Calculus to the Underprepared Students: Making History or Making Trouble? Pedagogy or Paternalism?" *UME Trends*, January 1993, pp. 1, 7.

2. Cohen, Marcus, Edward D. Gaughan, Arthur Knoebel, Douglas S. Kurtz, and David Pengelley, *Student Research Projects in Calculus*, Mathematical Association of America, 1991.

3. Corbett, G. R., "Minorities in Math—Two Programs that Work," *SIAM News*, July 1989, p. 1.

4. Dubinsky, Ed, and Keith Schwingendorf, *Calculus, Concepts and Computers,* Preliminary version, West Publishing Co., 1992.

5. Fusaro, B.A., and S.P. Gordon, "Filling the Tank: Models for PreCalculus Reform," *UME Trends*, August 1992, p. 8.

6. Hughes-Hallett, Deborah, Andrew Gleason, et. al., *Calculus*, Preliminary Edition, John Wiley & Sons, 1994.

7. Treisman, P. Uri, "Studying Students Studying Calculus: A Look at the Lives of Minority Students in College," *The College Mathematics Journal,* Vol. 23, No. 5 (1992), pp. 362–372.

Calculus Reform: How widespread is it? How major are the revisions? What are the issues?

James R. C. Leitzel
University of Nebraska, Lincoln

John A. Dossey
Illinois State University

National Science Foundation Calculus Program

The National Science Foundation initiated its Calculus program during fiscal year 1988. Since its inception, the program has been a cooperative effort between EHR's Division of Undergraduate Education and the Special Projects Office of the Division of Mathematical Sciences. During recent funding periods, EHR's Division of Research, Evaluation and Dissemination and the Division of Elementary, Secondary, and Informal Education have supported aspects of the calculus program.

Year	FY 88	FY 89	FY 90	FY 91	FY 92
No. Proposals	89	74	80	68	67
Awards	25	18	20	19	18

Table 1: Awards by year

For the 1992 fiscal year, the program title was changed to Curriculum Development in Mathematics: Calculus and the Bridge to Calculus. This reflected a broadening of the program to include consideration of the preparation for calculus and the interface between the student's senior year in high school and the first year of the college experience.

Not all of the awards involve curriculum development projects. Many of the initial awards were for planning projects, and several of the later awards were for workshops planned to disseminate the various reform efforts.

For example, in FY 88, the first year of awards, four were for full development projects that were funded for at least two subsequent years. One award went to the Joint Policy Board for Mathematics to begin the publication of UME Trends. The bulk of the awards, twenty, were planning grants. Included among the awards in later years was funding for MAA's publication, *Priming the Calculus Pump* [2], which appeared in 1990. This volume reported in detail on several of the calculus reform projects then underway and provided brief synopses of additional efforts.

The NSF initiative itself has moved from funding the development of new prototype courses to broad dissemination and implementation. As noted, during 1992, there was a program initiative to address the preparation for calculus. Other awards in 1992 were made to encourage large-scale implementation. One such award went to David Lovelock, University of Arizona, to implement a project over three years at Arizona State University, Brigham Young University, Northern Arizona University, Oklahoma State University, and the University of Arizona. A second phase of the project will include additional two- and four-year institutions and high schools in that geographic region.

Over time, there has been a winnowing of projects in terms of their acceptance in the broader mathematical community. Some initial awards went to colleges where the project was locally directed, and these had little transportability. At present, the most frequently cited efforts being implemented include the Harvard Consortium Project, Project CALC (Duke University), Oregon State Calculus Project, Five Colleges Calculus in Context, Calculus with Computing (University of Iowa), Calculus and *Mathematica* (University of Illinois), the Purdue Calculus Project, and Calculus from Graphical, Numerical, and Symbolic Points of View (St. Olaf College).

Information known at the present time (January 1993) indicates that the Harvard materials are being used at about 125 different sites, Project CALC at about 40

Highest Degree Offered	Doctorate	Masters	Bachelors	Associate	Total
Number	84	117	335	209	745
Percent of Total	11.3	15.7	45.0	28.0	100.0

Table 2: Description of Sample by Highest Mathematics Degree Offered

Change/Highest Degree	Doctorate	Masters	Bachelors	Associate	Percent
No Appreciable	34	45	133	115	43.9
Modest	39	65	151	79	44.8
Major	11	7	51	15	11.3
Percent	11.3	15.7	45.0	28.0	100.0

Table 3: Responses to Type of Change by Degree Type of Institution ($N = 745$)

schools, and the materials from Oregon State at about 40 colleges and 100 high schools. Of particular interest is the high retention rate of users. In many cases, schools trying reform materials in an experimental section one year, adopted and expanded the materials to all sections of calculus in the following year.

Each year since the 1986 Tulane Conference, there has been an increased number of sessions at national and regional meetings as well as conferences related to calculus and the introduction of technology in the teaching of calculus. Reform efforts are happening in many places and are continuing to penetrate the market.

With support from the National Science Foundation, the Mathematical Association of America (MAA) is conducting an assessment of the nationwide calculus reform effort. The task is to prepare a report for the mathematical community, and interested others, that indicates the current involvement of mathematical sciences departments (their faculty and students) in efforts to revise courses in calculus. The project is not intended to assess the outcomes of individual projects or to undertake an "official" review of NSF's calculus program initiatives, but to provide a report on the movement as a whole. The project has the acronym ACRE (Assessing Calculus Reform Efforts).

In Spring 1992, ACRE sent a one-page questionnaire to the chairs of all two- and four-year colleges and universities in the United States in an effort to get a quick measure of the status of calculus reform. This questionnaire requested departmental representatives to provide information concerning the status of calculus reform at their institution. A copy of the questionnaire is included as Appendix A.

Of the 2463 questionnaires mailed, 745 usable questionnaires (30.2%) were returned. These questionnaires were distributed across the various levels of institutions as shown in Table 1.

The responding institutions represent each of the fifty states, as well as the District of Columbia, Puerto Rico, and the Virgin Islands. The responses, shown in Table 2, indicate that at colleges and universities in almost all states there is at least a modest amount of change taking place in the calculus sequence.

Analysis of Responses

This preliminary report contains an analysis of the responses to the short questionnaire. Departments were asked to respond to a number of questions describing the change in calculus teaching and programs in their departments. The first of these questions asked:

To what extent has the calculus sequence in your institution changed over the past three years?

___ no appreciable change
___ modest change
___ major revision

Change/Highest Degree	Doctorate	Masters	Bachelors	Associate	Percent
Experimental	44	54	78	30	52.3
All Section	12	15	120	41	47.7
Percent	14.2	17.5	50.2	18.0	100.0

Table 4: Degree Type of Institutions by Calculus Reform Offering ($N = 394$)

Offering/Activity	No Appreciable	Modest Change	Major Revision	Percent
Experimental	50	143	13	52.2
All Sections	31	92	66	47.8
Percent	20.5	59.5	20.0	100.0

Table 5: Analysis of Offerings by Degree of Activity Reported ($N = 395$)

Table 3 provides the breakdown of responses to the question by degree-type institution. An inspection of the table shows that slightly more than half of the institutions responding are reporting modest to major changes in their calculus programs over the past three years. Only 11.3 percent are suggesting that their changes may be major in nature. Analysis of the data suggests no major differences in responses across levels of activities by any type of degree program.

A second question asked:

Is your department actively involved in a project to modify its calculus offerings?

__ Yes __ No

Of the 751 responses to this question, there were 403 (53.7%) "Yes" and 348 (46.3%) "No." Again, a breakdown by type of degree granting institution did not show any significant effect on the level of reform in the departments, indicating an active consideration of change in calculus was currently underway at all levels.

Those responding "Yes" to question 2 were asked to describe whether the changes involved "experimental sections" or "all calculus offerings." The response here ($N = 395$) indicated an almost even break between the use of experimental sections (52.3%) and all sections (47.7%) as a route to reform.

A further analysis indicates a slightly higher propensity of Bachelor and Associate degree program institutions to offer their reform curriculum to all sections, while Doctoral and Master degree program institutions tend to offer their reform efforts in experimental sections only. This conclusion may well be an artifact of size of the institution and management of students in sequential programs. It may also be a result of the availability of technology and other related resources needed to mount such a curricular change with the ability to offer alternatives for students enrolled in special professional programs, e.g., engineering, while the reform efforts are being field tested.

Another view of the change in calculus programs is obtained by comparing the offering of reform sections by the level of change reported in the department. When this is done, the data in Table 5 result. Here we see a propensity for those institutions indicating No Appreciable or Modest Change to offer more experimental sections and those indicating Major Revision to indicate changes in all sections. The levels of reform activity reported are consistent with the reporting categories employed.

Other data collected by the form deal with the percentages of faculty involved in the reform effort. These data, across institutions, were omitted in many cases, so they must be treated with care. Table 6 contains estimates of the percent of faculty involved in reform efforts by level of reform activity.

Another measure of the source of activity is the degree to which institutions have received funding to pur-

Percent/Activity	No Appreciable	Modest Change	Major Revision	Total
0–20%	31	81	8	120
21–40%	14	64	10	88
41–60%	9	31	14	54
61–80%	6	21	10	37
81–100%	17	51	31	98
Total	77	248	72	397

Table 6: Percent of Faculty Involved in Revised Calculus by Activity Level ($N = 397$)

sue their activities in the calculus reform movement. Responses to the question "Have you received funding for your calculus reform effort?" accumulated by the presence of a calculus reform project in the department shows that 42.2% of the departments reporting a project received some sort of funding for that project ($N = 384$). Of these 163 institutions, 95, or 59%, received some form of NSF funding to support their calculus efforts.

Second survey

ACRE has undertaken a second survey of 50 departments where major or modest changes in the calculus program were indicated in response to the first general questionnaire. The schools were selected in a modified random way. Attempts were made to assure geographic distribution and to find some sites where implementation of some of the more frequently cited efforts were being tried. The survey is rather lengthy—about seven pages of questions requiring detailed response. The returns should enable us to give a sketch of the process involved in attempting to implement a reformed calculus experience. The survey asks the sites to provide information on students and faculty attitude, on the involvement of client disciplines in the process of developing the new approaches, and information regarding the professional development of faculty and others who participate in the teaching of the course. In addition, the sites are asked to provide samples of test materials so that ACRE can make some statement of changes under way with regard to student assessment. In addition, the 1992 survey has been redone, with 1037 responses (36% return rate). The data have not yet been analyzed.

Assessment

In an early review of the NSF Calculus Initiative, Lida K. Barrett and William Browder wrote [1]

> In spite of the innovative nature of the proposals, the contemplated changes are not in the overall content of the course, but rather in the expectation of better student understanding; in greater stress on application and links with other disciplines; in the utilization of numerical methods and computer techniques; and in encouraging a fresh approach to teaching.

This theme was echoed at a July 1992 NSF-hosted conference for individuals interested in assessment issues related to the calculus reform effort. Believing that assessment should be based on articulated goals, the persons attending that conference prepared the following list.

Goals for the Calculus

- The student should learn the fundamental concepts of calculus: the derivative as a measure of change and the integral as a measure of accumulation together with the fundamental theorem of calculus which provides the exciting and useful connection between the two.
- The student should learn to formulate problems from science, engineering, and other disciplines in mathematical terms, and to use the concepts and techniques of calculus to provide new and additional information about the original problem.

- (cultural) The student should become aware of the beauty of calculus and of the power it brings to bear on problems from many disciplines.

- (societal) The instructor should ensure that the instruction in the course does not construct artificial barriers to the participation of women and minorities.

Not everyone may agree with these goals, but they seem to summarize fairly well the thrusts of the individual project goal statements. Some of the main issues that need to be addressed are issues that center around assessing the effectiveness of individual projects in meeting their own stated goals.

Changing curriculum is much easier than changing teacher practice. Yet the goals for calculus listed above relate more closely to teaching practice than to curricular topics. The most critical ingredient in fostering continued efforts is to know what is working in calculus reform today. In particular, are the changes having a positive effect on recruiting and retaining students in mathematics and mathematics related courses? We need to know the conditions under which change is being made successfully and how those conditions can be put in place in other settings.

Questions

In the process of thinking about student preparation for a "reformed" calculus experience, the following questions may need to be addressed:

- What is the needed level of paper-and-pencil skills in algebra, trigonometry, etc., for students entering calculus courses?

- How will students with a previous background in mathematics involving graphing calculator approaches perform in a calculus course utilizing high-end computer technology? How does this compare with the performance of students who have not had such a background?

- Are current curricular changes at the high school level consistent with the expectations of preparation needed for a reformed calculus experience?

- Since there are various curricular approaches to a reformed calculus course, is there a universal preparation for such experiences?

- What implications for student placement in collegiate mathematics courses arise because of new calculus or pre-calculus courses at the secondary level?

- How do we effectively address change in the college curriculum to prepare secondary teachers of mathematics and future collegiate faculty for the new calculus or pre-calculus courses?

- Are the methods of assessment of student learning consistent with the expectations for student understandings?

- How much change can be accomplished in either calculus or pre-calculus courses and still meet the current expectations of client disciplines? Are client disciplines changing curriculum and teaching at the same rate as mathematics?

Change in collegiate mathematics is taking place. It is an exciting time, a challenging time, and a critical time for seeking answers to the myriad of questions such change raises.

References

1. Barrett, Lida K. and William Browder, "Reflections on the Calculus Initiative," *UME Trends,* Volume 1, No. 4, October 1989.

2. Tucker, Thomas W., ed., *Priming the Calculus Pump: Innovations and Resources,* MAA Notes, Number 17, MAA, Washington, DC, 1990.

Appendix A

Survey
Mathematical Sciences Departments or Programs

Name:
Institution:
Address:
Phone: e-mail:

1. To what extent has the calculus sequence in your institution changed over the past three years?
   — no appreciable change
   — modest change
   — major revision
2. Is your department actively involved in a project to modify its calculus offerings?
   — YES — NO
3. If YES, do the changes involve (check whatever applies)
   — experimental sections — all calculus offerings?
4. About how many faculty, graduate teaching associates, and students are participating in the revised calculus offerings?

Group	Faculty	GTA's	Students
Number			
Percent			

5. Are you collaborating with other institutions?
   — Universities — colleges
   — two-year colleges — high schools
6. Have you received funding for your calculus reform effort?
   — YES — NO

   If YES, check all groups form which funding has been received:
   — National Science Foundation
   — Other Federal Agency
   — State
   — College or university
   — Other (describe)

New Developments in Advanced Placement Calculus

John W. Kenelly*
Clemson University

John G. Harvey*
University of Wisconsin

Introduction

The College Board Advanced Placement (AP) calculus program is deeply involved in the movement to integrate technology into mathematics instruction; the changes made in that program will have a major influence on schools and colleges. In the United States, in any given year, there are about 700,000 students enrolled in calculus, and within that group there are over 100,000 students enrolled in AP calculus. These numbers alone make AP a "major player" in the calculus reform movement; that influence is amplified ten times over by the way that many high schools structure their college preparatory programs around AP requirements. The AP program's significant role in determining the school curriculum has been acknowledged and the subject of educational debate for some time. But now we can add to the debate the recognition that AP is having a growing influence on college programs as well. Unfortunately, AP's growth in importance has not been accompanied by a better understanding of the program; so, the goal of this paper is to discuss the program's decision making structure and the history of some of its most recent considerations regarding the required use of computing technology, through graphing calculators, on its examinations.

The Advanced Placement Program Overall

The Advanced Placement Program is a series of introductory college-level courses and corresponding examinations available to high school students. The courses are shaped by course descriptions published by the College Board; the College Board also offers specialized training to several thousand AP teachers each year at workshops. Currently, almost half of the nation's 21,000 high schools offer AP courses that are taught by almost 50,000 AP teachers.

The AP examinations are administered nationally every May. In 1992, 388,000 students took over 580,000 AP exams. Since the inception of the program in the 1950's, more that 3.8 million candidates have taken some 5.6 million AP exams. Advanced Placement Examinations are currently offered in 16 fields; 29 different exams are offered. Subject to final approval by the College Board of Trustees it is anticipated that an AP examination in statistics will be offered soon.

Students who take Advanced Placement courses are intellectually capable, personally motivated, and academically advanced. The courses they take in an AP program are usually the equivalent of first-year college courses. Because AP students are generally capable, motivated, and advanced in their studies, colleges want them in their programs. Because AP students are capable of performing college-level work, their parents want the most for their money and credit where credit is due. Most important, because AP students are a significant and valuable part of the future of this country, we owe it to ourselves to make sure that they are getting the best that is available in education.

The AP program is this country's only national system for beginning college-level instruction in high school classrooms and the only national system for evaluating that learning. For nearly 40 years the AP program has been designing college-level course descriptions for high schools, supporting the training of their teachers, and monitoring its courses and exams in order to meet collegiate requirements. Today, nearly 3,000 colleges acknowledge the value of the AP program by regularly granting advanced placement or credit, or both, to students who present qualifying grades on AP examinations. This dynamic, continuously evolving system is the perfect catalyst for instigating fundamental changes in

*The authors have a long, continuing association with both The College Board and Educational Testing Service. Nevertheless, the opinions expressed here are those of the authors and do not necessarily reflect those of the College Entrance Examination Board or Educational Testing Service. In addition, the opinions of the authors about the Technology Intensive Calculus for Advanced Placement (TICAP) Project are their own and do not necessarily reflect those of the National Science Foundation.

introductory college-level courses. It is a system that works, that is in place, and that is respected.

A Short History of the AP Program

In the past, the AP calculus program has responded and supported curriculum proposals from the mathematical community. For example, in the late 1960's when the MAA released the CUPM document General Curriculum for Mathematics in Colleges, the AP calculus program split the calculus program offerings into Calculus AB and Calculus BC to parallel the MAA's call for two beginning college courses called "Math 0" and "Math 1" [3]. When the influential Sloan conference, held at Tulane University in January 1986, initiated the current calculus reform movement; the curriculum content workshop at the conference essentially duplicated the AP Calculus AB syllabus as its topical outline for a first-semester calculus course [4].

More recently, educational leadership by the AP program is seen in the introduction, in 1984, of the AP computer science program. Many observers say that the AP computer science course syllabus was a significant factor in changing the emphasis in introductory college-level computing courses from programming techniques to computing concepts. Likewise, the structured language requirements of the AP computer science course syllabus radically changed the computing environment in secondary schools (i.e., Pascal over BASIC). However, in the intervening years, college courses have forged ahead, and secondary computer science programs are having to strive to catch up. Thus, we see a classic push-pull relationship between an AP program and the colleges that benefit from it.

The move to require the use of calculator technology in AP calculus may generate this same "see-saw pattern." Graphing calculators in the schools could produce a generation of students entering colleges and pressing for changes in mathematics instruction. The colleges could respond, recognize benefits of the move, and carry the crusade even further and faster than the schools can respond. The pace of reform in calculus challenges the AP program to make changes at a much faster pace than in the past.

What Is The College Board?

Few school, college, and university faculty members realize that the College Board is a membership organization. The Board's offices are in New York City, and its members are the schools and colleges of the nation. Its programs are developed and monitored by committees appointed from these member institutions. Someone on every member campus votes the intentions of that member institution at the College Board business meetings. In the past the organizational structure was "one member, one vote," but in a recent major restructuring, the organization has assigned to each member three votes—one academic, one in financial aid, and one in guidance and counseling.

Many think that the AP program and the other College Board programs are Educational Testing Service (ETS) activities and that everything is determined by ETS in Princeton, New Jersey. Educational Testing Service is a non-profit corporation that, under contract, does operational activities including test development, grading, reporting, and research for its clients. In each case the customer retains control of the program policy. This operating structure is very different from that of the American Council of Testing (ACT) where policy and production are within a single organization. Every organization answers to its trustees. At the College Board the trustees are elected by the schools and colleges; at ETS and ACT the trustees are determined by internal means.

The AP calculus program is developed and monitored by our mathematical colleagues. The faculty involvement in this important College Board program is reflected in a lapel button that many AP participants wear that says: We are the AP program. Many mathematicians active in national calculus reform efforts are past or present members of College Board committees; viz., Phillip C. Curtis, Jr., John Dossey, Deborah Tepper Haimo, Winfred Kaplan, Don Kreider, Bernie Madison, John Neff, Anita Solow, A. W. Tucker, Tom Tucker, Bert Waits and the authors. Their voices have been heard within the policy-making channels of the College Board, and they will continue to be heard in present and future debates.

The current AP calculus committee has four college members, Wade Ellis (West Valley College), Bernie Madison (University of Arkansas), Anita Solow (Grinnell College), and Tom Tucker (Colgate University). It has three school members, Dan Kennedy (The Baylor School, Chattanooga, TN), Steve Olson (Hingham High School, Hingham, MA), and Margery Windolph (Mt. Pleasant High School, Wilmington, DE). Dan Kennedy is the committee chair. Ray Cannon (Baylor University) is the chief faculty consultant at ETS; he serves in an ex-officio capacity on the AP calculus committee. Two test development specialist at ETS, Chan Jones and Jim Armstrong, work closely with the committee.

The AP calculus program looked at a wide variety of options for the use of technology in calculus instruction.

A task force was appointed to work with the AP Calculus Committee, position documents were prepared and a conference was held to discuss this single issue. A wide array of advisors were invited to attend and a full spectrum of options were explored. Computer utilization was ruled out for several reasons. It was not feasible to expect every AP high school to subscribe to a particular computer-based calculus program. The selection of the software and hardware would have put a "straitjacket" on program innovation and curriculum development. Access to computer labs after school hours would have been difficult, and class schedules would have severely limited utilization during the school day. The cost of equipment and programs would have created serious equity problems for individuals and schools. Just as the colleges and universities must broaden successful participation in calculus, the AP calculus program must reach a wide audience of rich and poor alike. The cost of computers and expensive software would have been a move in the wrong direction.

Fortunately, powerful graphics calculator were an effective alternative. Yes, selecting calculators left a few things to be desired, but the number of omitted functionalities is being reduced by every new calculator model that is introduced. Yes, selecting calculators made the study of calculus in the schools diverse and open to a variety of approaches—but many saw that as a advantage. Yes, selecting calculators created an immediate need for a national program of professional development for thousands of AP teachers—but many saw that as a challenge. Yes, moving AP calculus into the broad utilization of powerful technology would put substantial pressure on the colleges to be equally innovative—but many saw that as a way for the College Board to reaffirm its historical role in curriculum innovation.

The AP Program as a Reform Agent

Some perceive of Advanced Placement as a conservative, centrally controlled national program that locks the basic academic subjects into inflexible syllabi. This perception is inaccurate and out-of-date. In the 1950s and 1960s, the AP program reflected its Ivy League and preparatory school origins and had the philosophy that AP courses should be the "common denominator" of its user group. This was appropriate when the user group was a small set of select institutions.

But in the 1970s and 1980s the AP program expanded into a wide base of users that included thousands of regional colleges and thousands of public schools. Today about 81% of the AP candidates come from public schools, and 23% are from minority groups. This growth took place over two decades. However, during this period AP courses were in the middle of the academic road. That is, the AP program reflected the curriculum of that time.

Through the late 1980s there was pressure to change this middle-of-the-road policy, and modifications were made. But it was in 1990, when the AP leadership at the College Board and the Educational Testing Service changed, that the AP program articulated the objective of today's AP program: to reflect what is best in American higher education. In addition, that "best" is being determined by the leaders in each academic discipline.

With its size the AP program now affects many university programs, and it is starting to "determine" some parts of the curriculum. Fortunately, the AP program recognizes its curriculum determining role and is dedicated to meeting the program's leadership responsibilities. The AP program is no longer bound to its original philosophy of being the common denominator; it wants to use the way that it shapes the curriculum to support the efforts of academic leaders to reform and revitalize undergraduate education.

To illustrate the way that AP is a change agent, note that at Clemson University half of the entering freshmen successfully complete much of their study of introductory calculus in the AP program; thus we cannot say that the AP calculus syllabus follows what Clemson does in its calculus sequence—for half of our students, AP is the Clemson calculus sequence. This means that AP has to be responsible and to maintain the best curriculum. That curriculum cannot be avant garde, and it can't follow fads. But, that curriculum has to lead.

AP candidates represent some of the finest students that enter the colleges and universities. In these fiercely competitive times, colleges and universities have to address the needs of this sought after student population. There are two driving forces at work: (1) the colleges' desire for high ability students and (2) the increased cost of higher education. AP students now leave secondary schools with increasing amounts of college credits in hand—in many instances up to a semester's worth and in significant numbers, a full year. With tight budgets, their parents no longer ignore the economic implications of a college's AP policy, and that makes the program a powerful force in the nation's system of higher education. Good AP students are known to "shop" for colleges and universities that have attractive AP policies. At the same time, colleges and universities want to bring these motivated and able students to their institution. With these forces at play, the changes in the AP program will be noticed, and the AP program will have a profound positive influence on the nation's educational programs.

Calculator Use on the AP Calculus Examinations

This is the AP calculus program's second attempt to use calculators in its exams. The first attempt was in 1983–85 [2]. In 1983 scientific calculators were allowed, and after 1985 they were denied. The decision to allow scientific calculators on the calculus examination was made by faculty members who were members of the College Board's committees and councils; these committees and councils recommend policy for College Board programs, in general, and for the Advanced Placement Program, in particular. As does the AP calculus program, each College Board program has a committee of school and college faculty members who determine content specifications for that program and who prepare examination questions and review and approve the examinations. When there are several test development committees within a subject, as there are in mathematics (AP, SAT, Mathematics Achievement Levels I & II), an advisory committee is created to monitor program-wide interests and policy. In mathematics, the supervisory body is the Mathematical Sciences Advisory Committee (MSAC) [8].

In 1980, MSAC, recognizing the growing importance of calculators, directed the AP calculus committee to implement the use of calculators on their examinations. This directive was in the context of an overall plan for the eventual use of calculators on all of the College Board examinations in mathematics. An AP calculator policy was put in place for the 1983 examination. The policy implemented was that scientific calculators would be allowed, but that they would not be specifically required on any test question (i.e., a "calculator neutral" style was used in developing the examination). Statistics gathered regarding calculator use on the examinations and on the scores of students showed that the calculators distracted students—especially, the lower achieving students; i.e., the candidates' personally reported levels of calculator use were negatively correlated with their scores. So the AP calculus examinations returned to a "no calculators allowed" policy in 1986. Note that in these decisions, the trend of the mathematics community to recognize the importance of calculators was addressed in a program-wide statement made by a College Board advisory committee, and the details of implementing the policy were determined by an individual test development committee. Thus, current and future recommendations of the AP calculus committee are not independent of overall programmatic considerations, and calculus "moves" were, and are, tempered by overall program needs and the requirements of other subjects (e.g., statistics).

During the years the College Board has established calculator policies for all of its mathematics examinations [7]. Beginning in October, 1993, it is recommended that all students taking the PSAT/NMSQT examinations use a four-function, scientific, or graphing calculator on the test. Pocket organizers, "hand-held" minicomputers, laptop computers, or calculators with QWERTY keypads will not be allowed. The same policy will go into effect for the SAT in March, 1994. The Mathematics Achievement Test Level IIC will require a scientific or graphing calculator starting in May, 1994. For the 1993 and 1994 examinations, the AP calculus examinations will contain questions for which a nongraphing, nonprogrammable scientific calculator is necessary or advantageous. The College Board has announced that "A graphing calculator MAY be required on the Calculus Examinations" starting with the May 1995 examination.

(Added in proof: Subsequent to the presentation of this paper at the conference, the College Board announced that graphing calculators will be required on the May 1995 AP calculus examinations.)

The TICAP Project

In response to the changes in the AP calculus program, professional development opportunities were developed. The academic leaders in mathematics determined that the integration of technology into calculus instruction was sorely needed, and the AP program responded with a national effort to make those changes with an effective professional development program for over 5,000 calculus instructors. Change without effective delivery to the classroom instructor is vacuous; so, the College Board is creating change in calculus instruction by enabling a national core of teachers to deliver change.

This new program is the Technology Intensive Calculus for Advanced Placement (TICAP) project, a special cooperative effort between Clemson University, the College Board, and the Educational Testing Service. The project is funded by the National Science Foundation ($940,000) along with substantial support from the Hewlett Packard Company ($180,000) and Texas Instruments Incorporated ($300,000).

The TICAP project has four goals:

1. To increase and improve the conceptual and problem-solving content of existing calculus courses by developing instructional materials that take advantage of graphing calculator technologies.

2. To increase teacher competence through their participation in workshops delivered by colleagues especially trained in the effective use of graphing calculators in calculus and precalculus instruction.

3. To create incentives for the nation's educational systems to incorporate computing technologies into calculus courses.

4. To explore the ways that graphing calculators can help currently underrepresented groups to improve their access to and successful completion of the calculus course.

In 1992, at the end of the grading of the AP calculus examinations on the Clemson University campus, 236 high school, college, and university faculty members attended a 3-day summer institute in workshop leadership that prepared them to give a new series of College Board calculus workshops. The summer institute participants learned how to prepare AP teachers to use graphing calculators effectively in their AP calculus classes, and the group now constitutes a cadre of educators prepared to be teachers-of-teachers. The College Board workshops will, in turn, help prepare AP calculus teachers throughout the nation to use graphing calculator technology effectively in their calculus instruction. There will be comparable TICAP summer institutes in 1993 and 1994.

TICAP will operate from 1992 to 1995 and is sequenced to facilitate the College Board's planned implementation of the required use of graphing calculators on the AP calculus examinations beginning in 1995. The faculty who participate in the TICAP summer institutes are the messengers of new and improved methods of teaching mathematics. Because they are carrying their message to AP teachers across the country through the College Board's extensive network of workshops and teacher education programs, they represent a forceful mechanism for change in both our high schools and colleges. The effort represents the best of systemic change; that is, working with national programs to cause significant and comprehensive change of the kind desired and endorsed by the leaders in a content discipline.

Much of secondary mathematics is preparation for the calculus, and more and more college majors are adding the traditional science and engineering calculus barrier—"only successful calculus students enter here." The mathematics community has called for calculus to be a "pump and not a filter," but many of the resulting open access proposals merge technology into the traditional curriculum. If the AP program is to do this, then a nation of teachers must have specialized instruction and materials. The TICAP project will try to supply those needs [9].

In its first two years at different times, the TICAP project has had on its development and instructional team Benita Albert, Judy Broadwin, John Brunsting, Ray Cannon, Frank Demana, Wally Dodge, Tom Dick, Deborah Tepper Haimo, John Harvey, Deborah Hughes Hallett, Patricia Henry, Dan Kennedy, John Kenelly, Don Kreider, Kathy Layton, Steve Olson, Don Small, J. T. Sutcliffe, Tom Tucker, Bert Waits, and Paul Zorn. Most of the members of this team work closely with the AP calculus committee. You might note that the group includes the president and past president of the Mathematical Association of America, the past chair of the MAA's calculus reform committee, and college faculty members that are all holders of NSF calculus reform grants. The AP teachers are equally noteworthy. All are currently or were previously AP committee members; essentially all of them are Presidential Award winners.

In addition to the assurances that the project gives to the College Board that AP teachers will be able to prepare for technology-enriched instruction, the effort will have an impact on the instruction of mathematics in two other significant ways. First, schools will be motivated to improve their pre-calculus programs to prepare for significant changes in the AP program. Second, as these students, with technology-enhanced secondary instruction, enter post-secondary education they will naturally expect and be prepared to handle mathematics instruction that takes advantage of today's computers and calculators.

The mechanism by which the College Board can make systemic changes in all of the AP academic subjects is by transmitting those changes to and through the (approximately) 2,500 high school and college teachers who are selected every year to read and score the AP exams. These academic leaders, called faculty consultants, make up a national network of dedicated teachers who gather together every year on three college campuses around the country to score the exams. The cost of their transportation to and from, and their participation in, the examination evaluation process is covered by the AP program. The marginal cost of their participation in a TICAP-like program will include only their additional living expenses.

Every year, the College Board's regional offices give thousands of workshops around the country designed to teach new skills to AP teachers. This system of workshops will be used to transfer the skills acquired by faculty at the Examination Readings to the tens of thousands of AP teachers who attend College Board workshops. Once again, a system for accomplishing the objective of the TICAP-like project already exists.

Just as mathematics has "shown the way" to other disciplines with efforts like the NCTM *Curriculum and Evaluation Standards,* the College Board plans to use the successful mathematical model, TICAP, to meet its curriculum leadership responsibilities in the other 18 academic subjects in the AP program. Initial efforts are already underway in Economics, Computer Science, and Biology.

In addition, the College Board has a regular schedule of publications (over 400) designed to teach teachers how to guide their students more effectively. These publications will be a vehicle for the communication of innovative and current teaching methods and skills, and the marginal cost will be virtually nil.

Problems in Future Test Construction

Introducing the use of calculators on standardized tests creates many problems for the test developers; some of these problems were described and discussed at the Joint Symposium on the Use of Calculators on Standardized Testing [6] sponsored by the College Board and the Mathematical Association of America and by Harvey [5] in his survey of the research on the uses of calculators on tests. One serious problem may arise from the possible differences in the ways calculators are used by persons who initially learned mathematics without having calculators (i.e., the test developers) and by persons who have used calculators throughout their mathematics education (i.e., students taking tests). The problem that arises is that test developers may overlook ways in which calculators can be productively used and students will not. The result can be that what is tested is not the same as what the test developer intended to test. In this section of our paper we will look at one example in a simple calculator environment that shows how serious this problem may be.

At first glance, numerical integration might not appear to be of concern on multiple-choice examinations that have questions dealing with the evaluation of definite integrals. After all, numerical methods are certainly one of the many ways that students should be taught to solve definite integral problems, and calculators simply broaden the available approaches. However, the earlier AP Calculus tests did not allow calculators, and definite integral problems had to be solved by evaluating antiderivatives. The introduction of calculators changes the testing environment, and, more importantly, calculator availability changes the statistical base on which the standardized tests are linked.

Notice that the midpoint rule is especially robust for relatively smooth functions, and a clever student could recognize that a scientific calculator has an "accumulator," the sigma plus key, and a ready "incrementor," the "memory add" and "memory recall" keys. With simple key strokes, values of the domain variable can be maintained and incremented with the [M+] key and then recalled with the [M rcl] key to evaluate the integrand function by using the standard function keys on a scientific calculator keyboard. After calculating the value of the integrand at the value of the domain variable, the approximating sum can be updated with the [$\Sigma+$] key. The first author applied this technique to the five problems on the multiple-choice part of the 1988 AP Calculus examination [1] and determined each of the correct answer choices with a simple two-interval partition and the mid-point rule. For the definite integral over the interval $[A, B]$, that process involved about a half dozen keystrokes at each of the two evaluation points $(3A+B)/4$ and $(A+3B)/4$. Three of the questions had zero as the lower endpoint and that made the problems even easier. The keystrokes took less than one-eighth of the two minutes, on the average, that the candidate is allowed for each of the multiple-choice questions on an AP Calculus examination.

In the past, definite integral problems on an AP examination would have required the candidate to be proficient with topics like: (a) integration by substitution, (b) integration by parts, and (c) the definition of absolute value. In the calculator environment, these problems may simply test whether one knows how to find a small number of values of a function on a scientific calculator. These dramatic changes in what the item is testing may change the measurement statistics on a standardized test item and change the validity of equated score procedures. Thus, as this example shows, changing the computational environment of the test presents special difficulties to test developers.

(Added in proof: Subsequent to the presentation of this paper at the conference, the College Board announced a policy for the 1995 AP Calculus examination that provides that part of the multiple-choice examination will not permit the use of graphing calculators and part of the multiple-choice examination will permit the use of graphing calculators. Part of the justification of this policy was the preservation of statistical links with the earlier AP Calculus examinations.)

Conclusion

Our intent in this paper was to discuss the ways in which the College Board's Advanced Placement Program in calculus is changing to respond to the changes being made in calculus courses by the calculus reform movement and to the increasing use of calculator and computer technologies in calculus instruction. We have also discussed the way in which the TICAP Project is responding to the need to train AP calculus teachers so that they can effectively and appropriately use graphing calculators in their day-to-day instruction and so, when graphing calculators are permitted on AP calculus examinations, their students will be ready. Along the way we have given you a short history of the Advanced Placement Program and have described how the AP programs are influenced by those who teach the AP subjects in high schools, colleges, and universities. We have concluded with an example of one of the problems that confront test developers when calculators are used on tests; our example also shows how versatile even simple computational devices can be when our knowledge of their capabilities is effectively combined with a sound, conceptually-based knowledge of mathematics.

For more than 35 years the AP calculus program has been an effective way for students to receive quality instruction in calculus while still in high school. Many colleges and universities have recognized both the quality of the instruction that these students have received and the quality of the AP calculus tests that they have taken by giving advanced placement or college credit or both to those students who have high enough scores on the AP examination. We strongly endorse the intent of the leadership that now—and in the future—the AP program will be a curriculum leader and not a curriculum follower. We believe that this intent insures that the AP program in calculus will implement appropriate and forward-looking policies, including policies regarding the use of calculator and computer technologies, and that the resulting reformed calculus program will be one that will be enthusiastically received by both the leaders of the present calculus reform movement and by those who want AP to continue to give students high quality calculus instruction that is recognized by colleges and universities as "the kind of calculus instruction they wish they could give to every one of their students!"

References

1. College Board. (1989). *The Entire 1988 AP Calculus AB Examination and Key.* New York: College Entrance Examination Board.

2. Commission on Mathematics. (1959). *Appendices to the Report of the Commission on Mathematics.* New York: College Entrance Examination Board.

3. Committee on Undergraduate Program in Mathematics. (1965). *General Curriculum in Mathematics for Colleges.* Washington, DC: Mathematical Association of America.

4. Douglas, R. A. (Ed.) (1987). *Toward a Lean and Lively Calculus.* (MAA Notes Number 6). Washington, DC: Mathematical Association of America.

5. Harvey, J. G. (1992). Mathematics testing with calculators: Ransoming the hostages. In T. A. Romberg (Ed.), *Mathematics assessment and evaluation: Imperatives for mathematics educators.* Albany, NY: SUNY Press.

6. Kenelly, J. W. (1989). A historical perspective on calculator usage on standardized examinations in mathematics. In J. W. Kenelly (Ed.), *The Use of Calculators in the Standardized Testing of Mathematics.* (MAA Notes Number 12). New York & Washington, DC: The College Entrance Examination Board and the Mathematical Association of America.

7. Jones, C. (1990). The use of calculators on College Board mathematics examinations in the 1990's. In J. G. Harvey, F. D. Demana & B. K. Waits (Eds.), *Proceedings of the Third International Conference on Teaching Collegiate Mathematics: Coming of Age.* Reading, MA: Addison-Wesley.

8. Orrill, R. (1992, August). Advisor to the College Board: The Mathematical Sciences Advisory Committee. UME Trends, vol. 4, 5.

9. Tucker, T. (1993). *1992 TICAP Resource Manual.* Clemson, SC: The TICAP Project, Clemson University.

Part 2

Workshop Reports

Report of the Content Workshop

Introduction

This report addresses the philosophy and content of the course or courses that should serve as the preparation for a reform calculus course. As such, we do not explicitly attempt to discuss the preparation for other college level mathematics courses, such as discrete mathematics, statistics, and finite mathematics. As a result, the ideas that follow are focused on preparing students for calculus at the current place and time. The recommendations below both affirm the changes proposed by the NCTM *Standards* and represent a transition from high school mathematics to the new calculus courses.

The types of courses that we address are offered in high schools, two-year colleges, four-year colleges and universities. The students at different institutions vary in terms of level of mathematical maturity, mathematical preparation and skills, ability and motivation. There is great variation in the time allocated to the preparation for calculus: from several years in high school to as little as one semester in college. We have not tried to distinguish between a single "precalculus" course and the sequence of courses from algebra through precalculus that represents the full preparation for calculus. Consequently, individual recommendations may not necessarily apply to all institutions, but the overall principles enunciated should apply to all mathematics offerings.

Finally, what we have addressed here is preparation for reform calculus, not necessarily for traditional calculus. However, we believe that many of the recommendations apply to the preparation for traditional courses as well.

What are the content goals of the "new" calculus courses?

The new calculus reform projects involve a change in focus in what it is important to learn, what is taught, how it should be learned, and the role of technology in the learning process. While there are many different versions of reform calculus, most of them share many essential elements:

- The emphasis of calculus should be on the fundamental concepts of the subject, not on symbolic manipulation. In some of the courses, the traditional manipulation is relegated to work that is done by a computer algebra system such as *DERIVE*™ or *Mathematica*®; in others, the level of manipulation is reduced and replaced by an increased emphasis on conceptual understanding of the fundamental concepts of calculus and their applications. In none of the reform calculus projects is the emphasis on having students perform long lists of similar exercises to gain dexterity at finding limits, derivatives or integrals or to produce graphs of functions.
- The topics in calculus should be approached symbolically, graphically, numerically and in verbal and written form.
- Calculus courses should emphasize modeling the real world and providing experience with problem-solving.
- Appropriate technology should be available at all times for graphing, numerical computations, and symbolic manipulations.
- The new calculus courses place a different level of expectation on the part of the students. Students are expected to think and not just to perform routine operations. This expectation is reflected in exercises, projects, examinations, and written assignments.

Principles for the "New Precalculus Courses"

The principles enumerated above for the reform calculus courses have immediate implications for the course or courses that prepare students for calculus. If the level of manipulative skill required for calculus is significantly reduced, there is less need to develop those skills in the precursor courses. The time saved can be devoted to other topics or to other emphases that better reflect what will happen in calculus. If there is a greater emphasis in reform calculus on conceptual understanding and mathematical thinking, then we should prepare the students by emphasizing these in the precursor courses.

Several new precalculus projects are currently underway, primarily in the secondary schools. These projects also involve less paper-and-pencil manipulative algebraic and trigonometric work on the part of the students. As with their calculus counterparts, some of these projects have students perform very little algebraic manipulation; others cover the gamut of ideas, but do not have the students perform the full array of complex manipulations usually covered in traditional courses. In none of these projects are students expected to perform long sets of similar exercises for the express purpose of developing high levels of manipulative skills.

The Content Issues Workshop participants believe that most of the points made above for the new calculus should apply equally to the courses that prepare students for it. We also believe that, in the process of implementing these principles, many different versions of precalculus courses will emerge, much as many different versions of reform calculus courses have emerged. We feel that this diversity will be healthy as different philosophies and organizations of topics will reflect a variety of perspectives.

In general, the participants agreed on the following set of fundamental principles:

Courses designed to prepare students for the new calculus should:

- cover fewer topics, but each topic should be covered more thoroughly and with more emphasis on fundamental concepts.
- place less emphasis on complex manipulative skills.
- teach students to think and reason mathematically, not just to perform routine operations. This higher level of expectation should be reflected in exercises, projects, examinations, and written assignments.
- approach each topic symbolically, graphically, numerically, and in verbal and written form with the aim of helping students construct appropriate mental representations of mathematical concepts.
- emphasize modeling the real world and develop problem-solving skills.
- make use of all appropriate calculator and computer technologies for graphing, numerical computations and symbolic manipulation. The full power of technology should be introduced in the service of learning the mathematics.
- promote experimentation and conjecturing.
- provide a solid foundation in mathematics that prepares students to read and learn mathematical material at a comparable level on their own, and to apply what has been learned in new situations.
- simultaneously serve mathematics and the physical sciences, the biological sciences, the social sciences, and other fields. The mathematics included should be presented as an elegant, unified and powerful subject that describes processes in all of these areas.

Rationale for the Precalculus Principles

There was complete agreement that traditional precalculus courses focus far too heavily on the process of learning procedures and far too lightly on learning mathematical concepts and their applications. They also attempt to cover too many topics, often as a consequence of the focus on algebraic techniques. Thus, the workshop consensus is that the new courses should de-emphasize complex manipulation, should cover fewer topics, but should cover those topics more thoroughly and with more emphasis on fundamental mathematical concepts.

In the process, there was considerable discussion, both during the formal working group sessions and in informal conversation, regarding the extent to which these courses should de-emphasize manipulation. While no one spoke in favor of long lists of routine exercises, there was no agreement on how much time could be saved or, equivalently, what should be deleted from the arsenal of paper-and-pencil techniques that good students should now be expected to possess. The high school teachers in the group pointed out that the NCTM *Standards* already call for a move in this direction and that students coming out of the high schools beginning in the next few years will likely have a lower level of manipulative skill than universities presently expect of incoming students.

The manipulative skills most often discussed were factoring, operations with rational polynomial expressions, radicals and radical equations, and trigonometric identities. Any consideration of the issues, however, must include simplifications and the solving of equations, inequalities and systems.

The issue is enormously complex. Consider, for example, the problem of reducing an expression such as

$$\frac{\left(\dfrac{x^2+x-2}{x^2-3x+2}\right)}{\left(\dfrac{x^2+3x+2}{x^2-x-2}\right)}$$

The solution of this problem is almost exclusively an exercise in factoring. Students who do not see such problems will likely not be handicapped in any of the new calculus courses; many students who are forced to perform such manipulations are likely to be discouraged, one way or another, from ever going on to calculus. Further, as some of the high school teachers pointed out, the traditional full treatment of factoring takes at least three months out of their syllabus; much more mathematics could be done if a significant portion of that time could be saved.

Alternatively, consider

$$\frac{1}{x-1}+\frac{1}{x+2}.$$

For some, this was the kind of manipulation students should not be expected to perform by hand; the ability to perform it by machine and to check the results numerically using substitution would suffice. Others felt that students should be able to perform this addition by hand and employ similar techniques whenever decomposition into partial fractions was required. However, no one spoke in favor of students needing to combine, by hand, significantly more complicated expressions involving, for example, quadratic terms in the denominator.

As a third example, there was a question of how many trigonometric identities should be required. Some felt that the Pythagorean identity $\cos^2 x + \sin^2 x = 1$ would suffice. Others would be content with this and the addition formula for $\cos(x + y)$. Others felt that students should know these as well as a handful of additional formulas. No one believed that students should have to learn the full arsenal of identities, such as the formulas for products of trigonometric functions, that are in traditional courses.

In a related direction, some participants felt that students should have some experience proving trig identities analytically, though certainly not to the extent currently favored. Others did not believe that this was essential. Yet everyone felt that students should be able to check whether or not a statement was an identity, either numerically by substituting values for the variable, or graphically by comparing the graphs of the two supposedly equivalent expressions.

As a fourth example, no one felt that it was necessary to require students to solve radical equations with two radicals. The only time that this comes up subsequently is in the derivation of the formulas for the ellipse and hyperbola and that can be demonstrated by the instructor. Yet, everyone felt there is value is solving relatively simple radical equations with a single radical term as an illustration of the use of an inverse function.

In summary, the resolution of these questions was felt to be beyond the scope of a brief conference such as this one. Some felt that research into the precise manipulative skills required in the new calculus courses should be the first step. Others felt that opinions on this issue are not likely to be swayed by research of any kind and that only time would tell which manipulative skills will survive. Some felt that the proof would be in the new courses that are developed to prepare for the new calculus.

There was complete agreement that conceptual understanding is central and must not be lost in the process of learning procedures. Thus, a major focus for the new precalculus courses should be to teach students to think and reason mathematically, not just to perform routine operations. This higher level of expectation should be reflected in exercises, projects, examinations, and written assignments.

In order to achieve this higher level of conceptual understanding, the participants agreed that topics should be approached symbolically, graphically, numerically, and in verbal and written form with the aim of helping students construct appropriate mental representations of mathematical concepts. The consensus was that non-algebraic ways of understanding mathematics should be highlighted, and the meaning of each representation, be it algebraic, visual or numeric, should be emphasized throughout. Thus, it is critical that students learn to investigate symbolic problems graphically and vice versa and that numerical values should be used consistently. Further, the participants agree that we should emphasize to students how important it is to develop the judgment to choose the correct tool (technology or pencil-and-paper). Finally, the courses should emphasize connections among different mathematical ideas.

Another common thread in the discussions was the need to make these courses problem-driven in the sense of using the applications of mathematics as the motivating idea. Thus, the courses should emphasize modeling the real world and develop problem-solving skills. The applications chosen should be of interest to students, not just to the instructor or to experts in another discipline. Finally, the applications chosen should relate students' real-life experience to mathematics.

Much of the discussion on the need for new preparation for calculus hinged on the issue of technology. It has changed what should be taught and how it should be taught. Several participants described a new calculator that has 64K of memory (more than the original Apple II), a full implementation of a computer algebra system, and two expansion slots that each take a 2 meg card. While many in mathematics think that technology affects only them, consider the impact of an expansion card with a full interactive dictionary/thesaurus on the English curriculum; now that calculators are allowed on the SAT exam, think what this would do to the verbal part; consider that the 2 meg card could also contain all related verbal questions from previous SAT exams.

It was evident that no participant felt that technology should or could be ignored. Instead, the consensus was that the new courses should make use of all appropriate technology for graphing, numerical computations and symbolic manipulation. The only caution expressed was that technology should not become the focus of the course, but rather that it should still be a mathematics course. Thus, the full power of technology should be introduced in the service of learning the mathematics. The extent to which this is done is a matter of individual philosophy. As with the reform calculus projects, some efforts will undoubtedly focus heavily on technology; others will utilize technology in less evident ways.

One of the primary uses envisioned for technology in these courses is to promote experimentation and conjecturing. Technology should not be used merely to get answers. Rather, students should use it to explore mathematical ideas without becoming overwhelmed with the manipulative work. For example, what patterns could they discover among the resulting terms by expanding $\sin 2x$, $\sin 3x$, $\sin 4x, \ldots$ using a computer algebra system?

Another area discussed briefly by the participants was whether the new courses should be narrowly focused on preparing students for calculus or should be broader to prepare them for mathematics in general. While we agreed to adopt the former in order to meet our charge for the workshop, it was evident that most participants welcomed the idea of broadening the content considerably. Suggestions to include additional topics such as data analysis, probability and simulation, recursion and iteration, and matrix algebra and its applications arose repeatedly. Some participants indicated that such topics might even be introduced in these courses in the service of preparing students for calculus.

Finally, some general philosophy regarding the new courses emerged. The participants believed that these courses should provide a solid foundation in mathematics that prepares students to read and learn mathematical material at a comparable level on their own, and to apply what has been learned in new situations. In particular, they felt that the principle of covering fewer topics means that different students will undoubtedly have "missed" some topic that may surface in a subsequent course. However, if the preparatory course prepares them to learn independently, they should be able to pick up a textbook written at the comparable level and learn some mathematics on their own. They should also be able to carry the mathematical knowledge and cognitive processes they have learned in one course over into other courses, both in mathematics and in the client disciplines.

Further, the participants felt that the new courses should simultaneously serve mathematics and the physical sciences, the biological sciences, the social sciences and other fields. The mathematics should be presented as an elegant, unified and powerful subject that describes processes in all of these areas.

Finally, the participants believed that every course should be approached as if it were a terminal course, not merely as a course that is intended to prepare students for a subsequent course. The result will be far better courses that provide students with a truly enriching mathematical experience.

Some Specific Ideas and Suggestions

The following are some more specific ideas and suggestions that arose during the workshop deliberations on particular topics and points that should be included in designing and implementing reform precalculus courses.

1. The Role of Algebraic Manipulation
   (a) Reduce the level of complex manipulation throughout.
2. Geometry and Graphs
   (a) Areas, volumes and geometric figures
   (b) Windows and scale
   (c) End behavior, local versus global behavior of functions
   (d) The idea of the "complete" graph of a function
   (e) Some graphs must be done by hand
   (f) The effect of parameters on scale
3. Numerical Work
   (a) Number sense
   (b) Use "messy" numbers; real data
   (c) Estimation and approximation
   (d) Accuracy and significant digits
   (e) When do we need an exact answer?

4. Thinking and Reasoning
    (a) Experimentation and conjecture leading to the writing of insightful proofs (this should not be a major focus, but an on-going theme)
    (b) Graphical verification (e.g., algebraic or trigonometric identities)
    (c) Use of quantifiers
    (d) Distinguish among conjecture, proof, counterexample, hypothesis and conclusion

5. A Continuing Theme: Functions
    (a) Functions as Modeling the World
    (b) Functions given by data/graphs and formulas
        i. Linear functions, constant rate of change
        ii. Exponential functions, constant ratio; for example, the population of California, 1890–1960
        iii. Trigonometric functions as model of periodic phenomena; for example, time of sunrise as a function of date
        iv. Other applications discussed: projectile motion, stopping distance for a car
    (c) Mathematical Classes of Functions will probably include:
        i. Linear
        ii. Exponential (any base) (more emphasis than in traditional courses)
        iii. Logarithms
        Note: exponential and logarithmic functions should be used throughout the course.
        iv. Periodic functions
        v. Polynomial functions (less emphasis than before)
        vi. Rational functions (much less emphasis than before)
    (d) Characteristics of Functions that should be included wherever appropriate
        i. Zeros
        ii. Rates of change
        iii. Increasing, decreasing, concavity
        iv. Extrema
        v. Composition
        vi. Inverses (Note: inverse problems should be used throughout as a unifying principle)

6. Other Topics that might be included:
    (a) Probability and simulation
    (b) Data analysis
    (c) Recursion, iteration, sequences and limits
    (d) Algorithmic thinking
    (e) Matrices and systems of linear equations
    (f) Combinatorics and counting

Report prepared by Sheldon P. Gordon, Deborah Hughes Hallett, Arnold Ostebee, and Zalman Usiskin

Report of the Teaching Strategies Workshop

The Teaching Strategies Workshop met for three hours on Friday and for an additional hour and a half on Saturday morning. The Friday sessions were spent discussing the general features of teaching strategies, administrative support for quality instruction, and methods of developing student self-confidence and enhancing their view of the value of mathematics. The results of these discussions were drafted into a preliminary statement that served as a focus of the final meeting of the workshop group on Saturday morning.

Pedagogical Methods

If mathematics instruction is to reach the goals currently held for its future, secondary school teachers and collegiate mathematicians and mathematics educators must:

1. Characterize, and reward, that which constitutes effective teaching;

2. Allocate *time as a resource* to support involved faculty;

3. Shift more responsibility for real learning onto students;

4. Encourage strategic use of available technology whenever appropriate and possible;

5. Engage applications of mathematics in a variety of areas as a conduit for learning mathematics concepts, skills, procedures, and problem-solving strategies;

6. Emphasize the idea that effective communication of mathematics is essential in today's world;

7. Nourish problem-solving and concept learning by groups as well as individuals, assigning fewer rote exercises and a reasonable number of problems involving mathematical modeling concepts;

8. Increased support must be allocated to mathematics if meaningful learning and high-quality instruction are to take place.

The members of the workshop were in strong agreement that the rethinking of instructional strategies is at the heart of the educational reform. For years, the lecture method has been the predominant model for mathematics teaching. In recent years, the focus in instructional strategies has shifted from the actions of the teacher to the nature of learning and the role of the student. In particular, this rethinking of instructional strategies involves:

- structuring learning activities which elicit direct student participation in a wide variety of class structures ranging from cooperative learning groups to student participation in lecture sections;

- shifting the emphasis in instructional planning from the content to be covered as the sole planning concern to also including the nature of the learning environment in which that content is studied;

- fostering student thinking and perseverance in working on problems involving new ideas or novel contexts, particularly real applications;

- encouraging all involved in mathematics instruction to incorporate calculators, computers, and manipulatives in developing students' understanding and abilities to represent mathematical ideas;

- increasing expectations for students' abilities to read, write, and speak about mathematics;

- supporting instructional methods which encourage the further study and learning of mathematics by all—including traditionally underrepresented groups.

Many felt that this rethinking of teaching strategies for collegiate mathematics was not limited to the precalculus and calculus, but was largely a result of the overall changes brought on by reform in mathematics education. Several present indicated that changing teaching was a topic where secondary teachers, postsecondary mathematics educators, and research mathematicians should all be working together. The joint implementation of similar teaching methods could ease the transition for students as they move from the secondary school to post-secondary education. In particular, participants noted that many secondary schools have changed their instructional approaches much more rapidly than universities due to their move to implement the National Council of Teachers of Mathematics *Curriculum and Evaluation Standards for School Mathematics* and the *Professional Standards for Teaching Mathematics.* Many were skeptical about the knowledge of many college and university faculty of the contents of these documents and of the resulting changes in secondary mathematics education.

Participants noted that these changes are related to teachers' self-confidence and their willingness to share "control" of the discourse in their class activities with students. Such changes require that instructors observe students while they struggle together to formulate ideas, to structure problems, to select problem-solving heuristics, and to develop and describe the solutions to a wide variety of mathematical situations. Central to these efforts is the focus on the development and support of student thinking and perseverance at the tasks posed. The development of a teacher's facilitation skills for managing small-group and cooperative learning situations does not occur overnight. The paper in this volume by David Smith discusses the sequence of changes that students first encountering student-centered, rather than teacher-directed, classrooms go through. Similar progressions take place for teachers making the change as well. It is during such phases that teachers need the support of their peers and administrators. The time of the development of team-building and the emergence of an esprit de corps is a threatening time for both student and teacher, but a period that must be passed if the desired changes in learning and schooling are to occur.

Although the topic was not directly tied to teaching strategies, many of the participants felt that changes in teaching strategies would not occur until adequate recognition and support are given to quality teaching in mathematics classes. At both the high school and university, teachers should feel the support and backing needed to implement non-traditional instructional methods. University participants expressed a view that the lack of recognition of teaching relative to research activity will limit the number of young faculty that will take the chance to get heavily involved in curriculum reform. Central to this change must be the support faculty will need when they implement small or cooperative group methods, change their assessment programs, institute the use of technology, or find that their distribution of grades changes. Perhaps most of all, departments at both the secondary and university levels need ongoing professional development activities that feature joint school-college work on the improvement of teaching, assessment, and curriculum models for precalculus and calculus. In particular, workshop participants mentioned the need for weekly seminars involving both high school and college teachers in planning for and sharing the results of teaching attempts. Additionally, instructors at both levels need to participate in sharing their ideas while trying those of their colleagues. All need continued opportunities to work with and integrate new technology into their teaching repertoires. In a like manner, the development of student skills in teaching, reading, and writing about mathematics required that mathematics teachers form, institute, and evaluate new types of classroom activities. Such changes in teaching will require a valuing of the efforts and increased support by mathematics administrators.

It was particularly clear to workshop participants that the size of mathematics classes at the university and the teaching load of secondary teachers and community college teachers have to decrease if mathematics teachers are to be able to work effectively with students in small groups. Cooperative learning, increased communication assignments, and significant use of manipulatives will not thrive or reach their full potential in classes that do not invite or enable successful attempts at changing instruction in precalculus or calculus. University leaders from departmental to central administration levels must exert leadership to bring the size of introductory classes to sizes equivalent to those for English composition. If students are to have the opportunities to communicate in speaking and writing as a part of their mathematical learning experience, then class sizes must approach 25 student maximums for significant interaction and learning to take place. If the University of Michigan can make such a move, other campuses can also.

Such changes also require faculty time—time an already overloaded teacher can ill afford to spend. Those responsible for instructional programs in mathematics must see that the resource of faculty time is appropriately distributed to maximize teaching efforts. University mathematics departments should consider the development of centers for the improvement of undergraduate instruction in mathematics that focus on student learning, faculty abilities, and changing assessments and articulation between courses.

Several participants discussed successful practices involving cooperative learning on their campuses. In particular, they noted the central role it puts on students to think and take some responsibility for formulating and choosing strategies to solve ill-defined problems or to work with others in completing projects involving the description and solution of problems requiring up to a month to complete. This promoting of change in student approach to monitoring their own learning came up over and over again as an important feature in changing student attitudes and understanding. Many indicated that once students "turned the corner" in taking this responsibility, not every little point needed to be taught in class, as students could be expected to learn on their own.

Technology

Another topic receiving a significant amount of discussion was the use of technology in the teaching of precalculus and calculus. The consensus was that technology was, and continues to be, a major instigator of change. The need to provide teachers at both the precollege and college levels with appropriate training in the use of new calculators, computers, and other forms of technology continues to be a major challenge for all involved in reform in precalculus and calculus.

There was considerable agreement that the calculator and computer provide teachers with challenges in getting students and faculty involved in working with opportunities presented through content-rich problems and questions. The secondary faculty present asked what were universities going to do when the present group of students, adroit with graphing calculators, reach the calculus classes of a university faculty not oriented that way. They described cases in which very capable students had discontinued their study of mathematics due to the strict anti- technology stances taken by their calculus teachers at the university. Other secondary faculty noted that they give their students advice based in part on the attitudes toward change of various university departments.

The participants agreed that technology has opened the door to the consideration of many important things ignored in the past. Its introduction, and acceptance, by many university faculty has given many of them a rebirth of interest and activity level after years of teaching. However, technology itself is not the subject matter. It is the lens through which we can look at precalculus and calculus, motivate change in content, vary contexts, and work to maximize student insight and understanding. Technology now allows to be done that which in the past could only be hypothesized. It opens the door to previously unthinkable problems and situations.

Communication

Considerable discussion was also devoted to the role of the development of communication skills in the teaching of precalculus and calculus. Participants felt that the development of student communication skills was a vital component of the desired changes in students' approaches to learning and knowing. Many discussed methods they had employed. These ranged from having groups of students write their interpretations or questions and exchange them with another group, to narrowing situations initially so that students had to talk to each other and then gradually expanding assignments and tools to allow for a greater diversity in approach and expression, and to providing multi-week projects for groups of 3 to 4 students using problems from client disciplines and then inviting individuals from these areas in to critique and to expand on the students' solutions. Many noted that the development of problems and communication skills (reading, writing, speaking, modeling, ...) in these formats was also a good first-order approximation to doing mathematical research. Such activities are good in helping students reflect on their own mathematical self-concepts while building perseverance and reflection into their overall mathematical behavior.

Project or problem assignments requiring communication get the students involved in talking to each other, not just answering. One participant described how she has groups of students make presentations and asks the remaining students to write evaluations of their presentations and the modes of investigation they employed. In turn, she evaluates their remarks as part of her assessment of both the group's work and her judgement of the emerging understanding of those writing the evaluations. Another noted that she looks for students' ability to plan, to summarize their procedures, to describe the solutions they developed, and to write convincing arguments for their findings. Several mentioned the MAA Notes Series volumes 14 and 16 on writing by Donald Knuth and Andrew Sterrett and the recent Heinemann text by Joan Countryman.

Assessment

There was also considerable discussion of the changes necessary in assessment brought about by the changes in teaching methods. Two maxims emerged related to this. They were: *Just because the students are learning the material doesn't mean something is wrong.* and *It doesn't have to be hard to be good.* Many participants noted that as students become involved in their learning in an active way, they approach the material differently, perhaps individually for the first time, and ask different questions in a real attempt to understand. As a result, they begin to see what is important and not important in their own learning and their performance improves. As a result, grades tend to increase. This, in general, provides a source of discomfort to many teachers, especially at the university level. The spread of grades becomes less important and the shift is to the quality of learning. Testing assumes less of a role of sorting and gives way to assessments that inform the improvement of instruction and learning.

Most participants agreed that the assessment of students has become "fuzzier" as a result of some of these changes, but more sources of information were being factored into the day-to-day and summative assessments of students. Almost all who had made changes in their programs stated that today they were using a wider range of student performance and products in arriving at final evaluations than five years ago. Specifically, laboratory work, projects, presentations, portfolio and profile grading are alternatives to the standard derivation of grades solely from paper and pencil testing. Information from "dumb mistakes" can be replaced by moments of bright insight if both are given credence in assessing a student's performance. In general, there was a strong feeling that such practices have opened up understanding of students and even their own understanding of the conceptual difficulty levels of topics within courses they had often taught before.

The use of placement tests in colleges and universities was also identified as a strong deterrent to curricular change in high schools. Many suggested that collegiate mathematics faculty interested in reform at the high school or college need to become active in understanding what effect their institutions' placement programs are having on secondary curriculum and teaching, on students, and even on their own programs. The bottleneck problem of students being held up in remedial courses after having successfully completed a full four-year secondary program needs special attention. Some suggested that such students be placed at the higher level, acknowledging their previous study, but given additional tutoring and help in making it through the first and second courses at the university. For students at the borderline, such an approach acknowledges their previous work, their at-risk position, and moves to bolster their self-images and keep them moving forward. Some noted that the correlations between the placement tests and final examinations were artificially high because both focus on algorithmic skills and facts, while the real proof of the viability of the courses resides in the knowledge, the thinking habits developed in cooperative settings, and in problem-solving that usually do not appear on either the placement or the final examination.

Research

The final portion of discussion dealt with resources on instruction, assessment, and learning. A great deal of high-quality research currently exists about the learning and teaching of mathematics in secondary and collegiate settings. Teachers need to become more aware of it and its implications for their teaching. They need to examine their own teaching in light of findings and experiences and reflectively work to improve in practicing what Ernest Boyer calls the "scholarship of teaching." However, all agreed that there was a great need for more information and guidance for mathematics teachers struggling to increase their effectiveness in promoting learning in an era of rapid change.

These discussions resulted in two final recommendations from the teaching strategies group. These were that teachers of mathematics:

1. Assess students' learning in ways aligned with course and program goals that improve learning and inform teaching

2. Prepare instruction based on knowledge of how students learn and reflectively monitor both student growth and program effectiveness.

Report prepared by Donald Bushaw and John Dossey.

Report of the Institutional Context Working Group

The Working Group on Institutional Context was asked to develop recommendations for actions that could be taken to (a) remove institutional barriers and to (b) improve inter-institutional articulation on issues that arise in connection with calculus and precalculus reform.

Members of the working group identified a variety of sources of institutional barriers and articulation problems: mathematics departments; high school, college, and university administrations; client departments; students' attitudes; the AP examination; placement testing programs; and the evaluation of courses for transfer credit.

Barriers to Reform within Mathematics Departments

Within mathematics departments the main source of problems seemed to be mathematics faculty members—both those who are satisfied with the status quo and see no need for reform and those who are fearful or lacking in self-confidence. Members of the working group pointed out that an increasing number of mathematics faculty have positive attitudes about the changes in mathematics instruction that are needed to take into account the effects of new calculator and computer technologies and the changing ways mathematics is being used in the world but that these faculty members often feel isolated and are unwilling or incapable of playing a leadership role in reform efforts.

These observations led to a discussion of the lack of reward for the substantial investment of time and energy needed to make significant curricular change. Indeed it was noted that in many university mathematics departments, colleagues count such activity negatively when making decisions on salary, tenure, and promotion. There was virtual unanimity among the working group participants that it would be desirable for the group to endorse the broader view of what constitutes valuable scholarly activity contained in Ernest Boyer's Carnegie Foundation Report *Scholarship Reconsidered.*

Specifics of how to affirm the value of participation in efforts to improve mathematics education were discussed. It was recommended that the Mathematical Association of America continue to sponsor sessions on mathematics education at its national and sectional meetings, that the American Mathematical Society increase its support of consideration of educational issues, that the National Science Foundation maintain funding for projects intended to improve the mathematics curriculum and instruction, and that calculus and precalculus reform projects continue to be publicized. It was also suggested that faculty isolation would be eased by the inclusion of all faculty—high school as well as college and university—in e-mail networks.

In addition, working group members pointed out that if reform is to be instituted on a large scale, special efforts will be needed to prepare part-time and adjunct faculty and graduate assistants to teach what to them are unfamiliar courses in unfamiliar ways.

Barriers to Reform from High School, College, and University Administrations

Members of the working group pointed out that administrations often use mathematics departments as revenue generating centers ("cash cows") by keeping teaching loads heavy and classes large. Administrations are also unused to thinking of mathematics departments as needing resources such as special types of furniture and physical space, equipment, released time for faculty development, support for team teaching, and so forth. Working group members also noted that inflexible administrative structures, such as rigid time schedules and fixed requirements about the utilization of classroom space, often impede change.

It was suggested that attention be given to lobbying administrations actively on the importance of developing new attitudes about the support needed for mathematics instruction. Working group members noted that both English and foreign language departments have occasionally succeeded in persuading administrations to reduce class size in order to achieve well-defined educational goals. Perhaps mathematics departments could emulate their efforts.

It was observed that a few institutions have already made significant changes in how mathematics instruction is structured. Publicity about those changes and their successes might be helpful. Formal studies of the educational effectiveness of such changes are also needed and would be helpful. Cases were reported in which higher retention rates and additional enrollment in mathematics courses has made such changes cost effective.

Barriers to Reform from Client Departments

Working Group members expressed only a few concerns about barriers from client departments. They noted that in many cases these departments are enthusiastic about calculus and precalculus reform efforts, but that in some cases their courses need restructuring in order to take proper advantage of the new calculus and precalculus courses. The need for ongoing communication between those involved in mathematics reform and their client departments, both at the high school and college levels, was discussed.

Barriers to Reform from Students

There was general agreement that the main barrier raised by students to successful calculus and precalculus reform efforts is the discrepancy between student expectations and the reality of the new courses. Students whose previous mathematics courses have focused on mimicry and drill are often taken aback by reformed courses that emphasize creative problem solving and conceptual understanding and that require significant amounts of reading and writing.

In many cases students ultimately become very positive about the new courses, but in the interim the evaluations of those teaching these courses may suffer, whether the course be at the secondary or post-secondary level. Thus working group members expressed concern that evaluation of faculty teaching quality not be based solely on present administration and student evaluation instruments. Suggestions were made that faculty evaluations include peer evaluation, both of classroom performance and materials development, and that data should be collected about subsequent student attitudes and performance. For example, a case was cited where students polled in advanced mathematics courses expressed more positive attitudes about the calculus reform courses they had taken than they had when they were enrolled in those courses.

There was also concern that once students have become accustomed to the new courses, they may have difficulty making the transition to follow-up courses that may not build effectively on the cognitive skills they have developed and that may not offer the variety of experiences with applications, concept development, and group learning the new courses have led them to expect. Some participants reported needing to make special efforts to encourage students to "hang in there" when enrolled in such traditionally taught courses.

Barriers to Reform from the AP Exam

In his presentation to the conference as a whole, John Kenelly discussed the role of the AP calculus exams in the calculus reform movement.

His presentation was followed by considerable discussion, during which a number of very negative comments were made about the inhibiting effect the exams have on curricular experimentation. Within the Working Group on Institutional Context, however, the views expressed about the AP exams were more moderate. Some members said they were frustrated because they felt the rigid AP course syllabus prevented them from participating actively in the development of new calculus courses. But they appeared to see the benefits of involvement with the AP program as outweighing the disadvantages of leaving it entirely. They were also encouraged by Kenelly's description of the openness of the College Board to significant change in the exams over the long run.

There was agreement that for those wanting to experiment right away with new approaches to calculus at the high school level, it would be desirable to broaden collaboration between high schools and nearby colleges and universities engaged in reform efforts so that college credit could be arranged for at a local level. The universities of Syracuse and Connecticut were cited as places where such efforts had been successful.

Barriers to Reform from Placement Tests

Placement tests were seen as an important problem area in the movement to reform calculus and precalculus courses. Many of the reform programs are moving away from an emphasis on timed testing and multiple-choice tests, and most programs de-emphasize purely manipulative skills. Yet virtually all current placement tests are timed, multiple-choice, and focus on manipulation. There was concern that present placement tests do not measure the strengths of students emerging from the new courses and that as a result students would not be placed appropriately. In addition current placement exams would not appear to measure accurately the skills needed to succeed in some of the new courses. There was a general consensus of members of the working group that new placement procedures were needed. It was suggested that the group recommend that those involved with each calculus reform project carefully examine their materials to determine exactly what prerequisite knowledge and cognitive skills are needed, that there be cooperation among the various reform projects to determine commonalities among prerequisites, and that the MAA be encouraged to play a leading role in developing new placement procedures.

Barriers to Reform from Evaluation of Courses for Transfer Credit

Some members of the working group noted that as the variety of different calculus and precalculus courses increase, it will be more difficult to determine how to assign transfer credit. Other members responded that departments such as English have faced and resolved similar difficulties: that is, that an introductory writing or literature course at one college or university may be very different from a similarly titled course at another. Even within a given department the content of a single course may vary dramatically from one instructor to another. Thus it should be possible to assign transfer credit to the new courses much as is done now.

Of course, since mathematics courses at one level often build on specific knowledge from previous levels, the details of content cannot be ignored entirely. However, it was pointed out that content issues can generally be resolved satisfactorily and even easily on a local level; for instance, in situations where certain community colleges send large numbers of students to nearby four-year institutions. But in such situations, it was noted, it is important that people who genuinely understand mathematics be involved in drawing up articulation agreements.

The Recommendations of the Working Group

The working group met for three sessions. In the first two, issues were identified and suggestions for recommendations were proposed. These were written up and presented in a plenary conference session. Then the working group met again to discuss revisions, and there was additional informal discussion about the recommendations both inside and outside the working group. A presentation was made to the full conference at its final meeting, at which time a few additional revisions to the recommendations were made. At this final meeting, however, a decision was made not to take formal votes on working group recommendations but to write up a summary of the discussion of the issues and the recommendations that were proposed. The following statements, therefore, represent an informal consensus of opinion.

1. Departments of mathematics and the institutions that house them should adopt the broader view of what constitutes valued mathematical activity that is described in Ernest Boyer's Carnegie Foundation report, *Scholarship Reconsidered*.

2. Colleges and universities need to provide the resources for educating all faculty about the new developments in mathematics education that are necessary to successfully implement curricular reforms. Among these developments are cooperative learning, uses of calculator and computer technologies in instruction, new assessment techniques, and awareness of the variety of learning styles that exist among students. Further, there is a need for more diverse methods of judging effective teaching beyond numerical summaries of student evaluations.

3. Increased attention should be given to preparing undergraduate and graduate students to be teachers and to fostering the development of new and part-time faculty as teachers.

4. Positive involvement in innovative projects in mathematics teaching and curriculum development should be rewarded and recognized, especially in tenure and promotion decisions.

5. An organized effort should be initiated to gather and publicize information about the various precalculus and calculus reform projects including demographic data, longitudinal studies of student attitudes and achievement, and studies on the nature of student learning and the effectiveness of new instructional strategies.

6. Mathematics instruction in high schools, colleges, and universities should be restructured to enable successful implementation of precalculus and calculus reforms. Specifically, class sizes should be re-

duced, teaching loads lowered, and needed equipment, furniture, and physical space provided. To help convince administrations to devote the funds needed, we recommend sharing information about departments that have restructured successfully.

7. Calculus reform projects should singly and jointly formulate a common set of prerequisite experiences and cognitive skills which will become the basis for new placement procedures. The Mathematical Association of America should play a leading role in developing such procedures and colleges and universities must devote the necessary resources needed to their own placement programs.

8. Increased cooperation among institutions is needed regarding articulation and transfer of credit. State universities should take the lead in developing articulation agreements with area community colleges and high schools and should involve representatives from the respective mathematics departments in the process. College credit should be given for calculus reform courses taught in high schools through cooperative arrangements with local colleges and universities, e.g., the models at the University of Connecticut and Syracuse University.

9. The faculty of all secondary and post-secondary institutions should have access to e-mail so that faculty will have the opportunity to communicate directly with colleagues all over the country to exchange ideas, materials, and so forth.

Report prepared by Susanna Epp and Lee Yunker.

Report of the Course Context Working Group

Four major themes organize this report on the discussion of the working group on course context: the calculus course in transition; implications of changes in content and approach for programs that follow calculus; preparation for the "new calculus" courses; and improving access to calculus.

The working group used a liberal definition of course context to prompt discussion about the content and spirit of the new calculus courses. Many times the discussion moved away from issues of calculus topics and course goals in the context of an educational program to institutional and cultural impediments to mathematics reform. The reader will recognize many of the issues in this section as concerns expressed from different perspectives in the other working groups.

The Reform Calculus Courses

The new calculus courses advocate revisions in content, and feature different styles of learning and doing mathematics. In reviewing the possible effects of calculus reform on postcalculus mathematics and client disciplines, the working group wrestled with the definition of topics that might characterize a new calculus.

Mathematics departments are faced with developing calculus for certification as well as for later mathematics. Calculus is a required course for business, engineering, and medical programs. In some schools, the same calculus course that serves the needs of other departments is the course that launches mathematics majors in course sequences that lead to a mathematics degree. Therefore, if we were to do a general calculus, what has to be in it to serve well the students who will be majors? How can it best serve the students who will use math in other areas?

The working group agreed that a contemporary calculus would be built upon the core concepts of limit, derivative, and integral. That triumvirate of ideas is not sufficient to provide a curriculum distinct from one that might have been in place 10 years or 50 years ago. However, the new calculus curricula challenge tradition in assumptions about the student as learner and in the nature of mathematics in general education. These assumptions center on the meaning of mathematical maturity, the importance of student initiative in building mathematical ideas, and the role of applications in calculus.

The working group considered different ways a student could reach understanding of mathematical concepts and demonstrate that understanding through means other than proof. They asked whether proof was the "asking of reasons or justification" to explain what you are doing, and whether it is better to use the words "mathematical argument and generalization" in discussions of mathematical reasoning. What numerical or graphical techniques provide students with the knowledge of great ideas of calculus, and can they do a better job than formal proof? There are certainly great theorems of the calculus that may be too obscure for many students when developed via proof. For example, the mean value theorem is frequently presented in a formal development that neither convinces nor links to immediate uses for the theorem. However, the theorem is obvious graphically. It can be illustrated in a way that expands a student's knowledge for presenting evidence of mathematical relationships via graphs. Development of limits may be an topic well suited to the numerical and graphical techniques now available through popular technologies. For example, it is possible to use a graphing calculator or a spreadsheet to demonstrate that $\lim_{x\to 0} \frac{\sin x}{x} = 1$. Hand-held graphing calculators permit the student to see how the graphs of $y = \sin x$ and $y = x$ are very, very close to each other as the student zooms in on the graphs near the origin. A spreadsheet can show numerically how numerator and denominator each approach 0, but the quotient approaches 1. A column could be constructed starting at 1 and using the recursive step $a_n = \frac{1}{2}a_{n-1}$. The second column would be $\sin a_n$, and a third column would be the quotient of the corresponding elements in the first two rows. Popular spreadsheets perform computations with many significant digits of accuracy, so they offer the student an opportunity to study small differences very close to zero.

Despite the success of the new calculus in presenting mathematical ideas through different representations and activities, members of the working group expressed a concern that important views of mathematics that emerge from traditional courses should not be lost. Mathematics does some things very well. There is a feeling in mathematics that when you get an answer, you have ways of knowing it is correct. We should point that out to our students, and we should show the connections and uses in other areas. That becomes a goal for the teacher of calculus that may transcend questions about textbooks. A good teacher will be explicit in helping students organize the calculus about great ideas.

The broader view of calculus represented by the new texts will serve the mathematics major as well as the liberal arts student. Many post-calculus courses require three semesters of calculus. The stated reason is that such preparation evidences mathematical maturity. This means we must ensure that the calculus is focused on concepts and not just three semesters of symbol manipulation.

A common theme of reform movements in mathematics is that the student must be an active participant in learning. The new calculus courses feature a variety of techniques in presenting mathematical ideas. The quality of the writing of lessons and explanations in the new texts shows that the authors believe that students can learn from reading. It is recommended that students write extensively, both as a way to reach understanding and as a way to demonstrate mathematical reasoning. Continued mathematical growth comes from having students explain in writing what they are doing and why they are doing it. Technology becomes more than a window through which one views mathematical phenomena, but also a tool by which the student can perform mathematics and explore relationships. Good applications and problems promote mathematical communication among students. There was a strong feeling that the calculus reforms need to be extended to other mathematics courses. Students need to be engaged as active problem solvers not only in reform courses but in all mathematics courses. However, there was also concern that calculus teachers shouldn't be focused exclusively on avoiding lecture. There is too much math to be discovered by every student.

The new calculus texts incorporate many applications. Mathematics is strongly positioned to take advantage of the opportunity to connect to other disciplines and to the real world. There are attempts to develop good applications from other subject areas, even in traditional calculus courses. This brought the discussion to the role of calculus courses in the curriculum of majors outside of mathematics. The working group expressed strong feelings that the lessons of the new calculus should be applied to engineering, medicine, and business departments. Client departments must reform as well. Are they really using the calculus in their courses? Or are they finessing it? However, these questions caused some soul searching by the working group. Can mathematics teachers present applications really well? Do they know enough about the applications to link the calculus in an accurate and compelling manner? Part of the issue of applications is jargon: 'marginal' related to derivative is recently adopted from the field of economics and may not be in the mathematics vocabulary of the calculus instructor. Nevertheless, popular applications such as predator-prey, AIDS epidemic models, economic relationships, and population models illustrate major ideas of calculus well and should be expected to be within the experience of the calculus teacher. We recommend that mathematics faculty build connections with other departments to maintain expertise in applications. The client department can help mathematics professors incorporate realistic and current applications into mathematics courses. The mathematics faculty can serve as messengers of reform who may help other departments in shaping improvements in related programs and courses.

Articulation is particularly important when other disciplines provide calculus instruction. We must scknowledge the reasons other departments have their own calculus. One reason given by other departments is that the math faculty don't care about applications, and don't know good applications. Some other views attributed to folks outside mathematics as justifications for discipline-based calculus courses include, "We could teach in one course what the math department teaches in five," and "the math department can't shut up." These attitudes may have come from bad experiences with mathematics departments twenty years earlier and are not applicable to the reform calculus courses.

But is the problem with the existing course that the math department tried to accommodate too many people? We need to get our own house in order. We need to define the body of knowledge students should have as they move through our courses. We should be asking whether the course makes sense, has the correct content, and whether the students are engaged in learning. The math department should be the major player in deciding what goes into our courses. However, this brings to the fore a critical issue on the role of the individual teacher: should an instructor be permitted to define his/her own course? This is a serious problem for schools that use adjunct or part-time faculty to teach mathematics courses. The new calculus courses define content and spirit of mathematics that require coherence among courses and mutual support among mathematics faculty.

Some other disciplines provide broad survey courses that serve as a component of liberal education as well as the initial stage of training in that subject. The members of the working group challenge mathematics departments to keep more calculus students in mathematics. This requires the linking of improvements in courses that follow the calculus to the best content and pedagogy of the new calculus courses.

Programs which Follow Calculus

Technology can be a stimulus to reform in later mathematics courses that may be later courses in mathematics or courses in client departments. This technology can include all-purpose mathematical software, such as *Mathematica* or *Derive,* that might get repeated use from course to course, or software to teach a specific concept or solve a single problem. Students are more fluid in their use of different software than are faculty, but one must balance the overhead of learning a particular piece of software with its long-term utility.

Courses that come after calculus must capitalize on the students' prior experiences. They should have some obvious basis in the activities and goals of a reform calculus. Calculus teachers can establish connections to instructors in post calculus courses in mathematics and courses in other subject areas and articulate the changes in mathematical experience the new students will demonstrate.

Courses that follow calculus should have applications as well as formal mathematical proof. There should not be an abrupt break, a chasm created where on one side there is a course rich in applications and on the other there is a course rich in theory. Both applications (to other disciplines and the world as well as to mathematics) and concise convincing arguments (proofs) should be a part of any good mathematics course. Courses preceding calculus and courses that follow should require students to engage in justifying their work and to learn how to produce convincing arguments.

We urge instructors to attend to the egress from reform calculus classes. If we view the calculus course as a source of future mathematics majors, we should do some market analysis. It might be good to ask mathematics graduate students when they became math majors and why.

Courses which precede the reform courses

It was easier to prescribe good instruction for courses before calculus than it was to define prerequisite content. The working group started with an axiom that "what is good for the students in the reform calculus courses, should be good for students in the precalculus course." A natural recommendation was for schools to practice the National Council of Teachers of Mathematics *Standards for Curriculum and Instruction.* Student characteristics or habits that might emerge from education that would follow the *Standards* would include an ability to explain orally or in writing the steps of a solution or a mathematical argument; facility with technology; and experience of working in groups. The student should have some tolerance for delayed gratification.

The content issues related to functions. The first semester of calculus is a survey of single variable functions. The more experience a student has with functions, the better. This requires a shift from the dominance of algebra manipulation and trigonometry identities that dominate precalculus curriculum. But if the calculus course is concept-based, then it should be easy to make predecessor courses concept-based. Students must be able to understand what a function is. They should be able to take a description of a problem and see where a function resides in the problem. The use of functions to model data requires that students come to the calculus with a repertoire of functions. For example, students should be familiar with the functions on a scientific calculator. They should be able to recognize the graphs of these functions, be able to write algebraic expressions for these functions, work with tabular and numerical representations of functions, and recognize a significant application of each function. The students should be able to graph functions and use that visual representation to formulate, interpret and solve problems.

Graphing technologies offer a way students can answer questions that have previously been restricted to the calculus. For example, what are the maximum or minimum points for a function? Where does the graph indicate changes in rates? What is the average value of the function over a specific interval? Each of these questions is best addressed through problems that are in the context of an application. The study of these issues in precalculus courses does not preclude them as content for the calculus, but rather offers students an opportunity to prepare for success in the calculus course.

Some members of the working group challenged precalculus as an organizing theme for the course before calculus. The term 'precalculus' is pernicious. A fourth year high school mathematics course should be a capstone course. It might have goals that are broader than just preparation for calculus, i.e., experience with statistics, computing, or discrete mathematics. A better name than "precalculus" might be invented for this course.

The corresponding course at the college level might very well be called precalculus if its intent is piping a larger number of students into a successful experience in calculus. However, there is unhappy evidence that precalculus at the college level isn't a solution for increasing calculus success. At some colleges and universities, very few people who take precalculus survive the first semester of calculus with C or better. Therefore, should there be a course called 'precalculus' in college that has its only role preparation for calculus? Although members of the working group indicated that the answer to this question was "yes," they noted lack of enthusiasm for such a course. It may be more important to base a course before calculus on a broader curriculum, just as was recommended for a high school "capstone" course. Whether they be high school or college, precalculus courses must focus on the skills of learning as well as the skills of content.

Improving Access to Calculus Courses

Almost every discussion of the working group converged on the need to promote higher enrollment in calculus. Representatives of all types of institutions expressed conviction that policies and procedures of mathematics departments could be altered to encourage students to take calculus. We recommend that departments clarify the content and difficulty of courses, remove barriers to progress in mathematics, provide new ways in which students can show readiness for calculus, and facilitate reentry into mathematics for older students.

Every mathematics department should provide articulate and accurate statements of what a course is going to offer, how difficult it will be, how it will compare to other courses, and how it will engage students. This description can be developed by the instructor. Asking students in the course "what would you have liked to know about the course before you registered for it," or "what mathematics or learning skills do you think someone should have for this course," might provide an interesting start on a course description. This is different from the action "write down the skills you should have in precalculus." Many students in traditional courses learn to do things from templates. We need students who have learned the content of precalculus in flexible ways. Formal prerequisites have been overrated. Maybe we think that students need to know more than they really do to succeed in calculus. On the other hand, maybe there are some content issues, like functions and understanding the algebra of the line, that can't be remedied in the time allowed for a calculus course. Should we try to define prerequisites in terms of "mathematical methods?" This might be "use variables, develop equations as models of phenomena, understand functional relationships." Nevertheless, there were advocates for the premise that the calculus teacher should be able to teach all the necessary skills and understandings as part of the calculus course.

Departments should expand their criteria for placement in calculus. If the reform movement in calculus stimulates change in precalculus instruction, then many students should possess skills and habits that lead to success in calculus. However, these might not be the skills and habits currently measured through conventional placement tests and interviews. Placement should be inclusive. In this sense, the success of a course called precalculus should be demonstrable. Precalculus should not be an excuse for the gratuitous inclusion of courses as barriers to the calculus. Therefore mathematics departments must reexamine the prerequisite courses and the process by which students must wend their way to calculus.

We recommend that departments facilitate reentry into mathematics. There are many students who should take calculus but were steered away from four years of mathematics in high school. Substantial numbers of students have marginal grades from high school courses but might blossom in a vigorous course. Community colleges can attest to the number of nontraditional students entering post-secondary education. Maybe these students should have some different ways of getting up-to-speed for calculus.

College placement for mathematics should be advisory, not mandatory. The members of the working group were very concerned with the mindless enforcement of selection procedures practiced at many colleges and universities. Even if tests and scores could give a first approximation to an appropriate course for a student, there should be methods for additional information to improve the recommendation. When all is done, the student should be able to make his/her own choice. This prompted discussion of the need for departments to examine their methods of assessment in all courses. How calculus knowledge is tested and graded may be forcing students away from mathematics rather than giving students opportunities to show knowledge and learning.

Recommendations

1. The innovations of the reform calculus courses promote new views of mathematical maturity, emphasize the student as active learner, and rely on applications to motivate and validate concepts of calculus. These views may run counter to the model of calculus teaching held by mathematics instructors

and against the traditional content expectations of client disciplines. A first step for every mathematics department is a review of whether their calculus course makes sense, whether the content is right, and whether the students are engaged in learning. The department must define goals for student learning and growth, and enlist all of its members in a coherent effort to reach those goals. Since the calculus course should not exist in isolation, the department should engage in vigorous articulation with client disciplines about the skills and habits of mathematics that have most worth for their students.

2. The issue of abstraction and mathematical argument in calculus was one that dominated much of our discussion. We recommend that reasoning and justification techniques be viewed in the context of student growth in mathematics. The calculus should not be imposed with the full responsibility for development of mathematical formalism expected of the mathematics major. The calculus, however, should require students to grow in their abilities to communicate mathematical ideas in clear and concise fashion and to demonstrate understanding of concepts through graphical, numerical, verbal, and symbolic expressions.

3. The general themes of the reform movements should be extended to courses that follow the calculus and those that precede it. These require that the student be an active participant in the learning process, that the faculty employ a variety of instructional styles, that technology be integrated into the content and methods of mathematics courses, that concepts be linked to applications, that students be encouraged to develop strong communication skills (writing, speaking, and reading), and that each course be viewed as incrementing the level of abstraction and mathematical argument.

4. Preoccupation with calculus preparation must not monopolize predecessor courses in the college or high school. There are many mathematical ideas that prepare students for later mathematics or for the mathematical thought that has utility for other disciplines. Courses should be based on activities and content that can be shown to have value to students in their acquisition of mathematical power.

5. The recognition of broader goals for the courses before calculus implies that placement of students must recognize a wider range of skills, knowledge, and motivation than is now measured on a typical department placement test. We recommend that mathematics departments review placement procedures from the standpoint of their being barriers to calculus. Mathematics faculty must devise ways to facilitate reentry into mathematics for underrepresented minorities and for students who may have been steered away from four years of mathematics in high school.

Report prepared by Carolyn Mahoney and John McConnell.

Part 3

Contributed Papers

Calculus Papers

The Rising Tide: Supercalculators in Calculus

Donald R. LaTorre
Clemson University

We are now over six years beyond the introduction of the first sophisticated graphics calculator (the HP-28C in January 1987), and there is no longer a debate about the appropriateness of such devices in collegiate mathematics. These units, especially the higher level Hewlett Packard HP-48 and Texas Instruments TI-85 calculators, have proven themselves to be effective and valuable tools and are no longer regarded as being inferior to microcomputers in their ability to help bring about substantial change in the way students learn mathematics.

Over the next few years our nation's colleges and universities will be admitting the first wave of the future generations of high-school students having substantial experience in learning mathematics with the aid of programmable graphics calculators. There are several powerful forces that are pushing this tide.

The National Council of Teachers of Mathematics (NCTM) *Standards* recommend graphics calculators in every school mathematics classroom and this expectation is being rapidly embraced by schools all across the country. As a result, students are buying graphics calculators in record numbers: some estimates are that some 6,000,000 units will be sold in 1994 alone. The College Board's new policies are also at play. In October 1993 they allowed any four-function, scientific or graphics calculator to be used on the PSAT exam, and in March 1994 this policy was extended to the new SAT tests. You can thus expect parents everywhere to be purchasing units for their teenage children. The move toward graphics calculators is being further accelerated by the College Board's anticipated plans to require the use of graphics calculators on the 1995 Advanced Placement calculus examination. The tide is indeed rising!

Clemson University is currently at the crest of this tide with its established program that requires high-level graphics calculators (the HP-48) throughout the first two years of mathematics for students in science and engineering. In this article we focus on the following two questions: how has Clemson managed to implement the highest level of calculator technology; and what are some of the lessons to be learned from the Clemson experience?

Large Scale Implementation: A history and some lessons

During the 1992–1993 academic year, we taught 56 classes in single and multivariable calculus, differential equations, and linear algebra in which every student had his or her own HP-48 calculator. This was the third year that Clemson required the advanced symbol manipulating units in almost 50% of its offerings of these classes during the regular academic year. With class sizes averaging 36 per class, our experience is both extensive and broad-based. And a decision in March 1993 of the mathematical sciences department required that all students in all sections of calculus, both single variable and multivariable, have their own HP-48G or -48GX supercalculators by the summer of 1994. The 1993–1994 year was one of transition from the 50% level to full implementation of hand-held technology; we maintained an appropriate number of noncalculator sections of Calculus II and Calculus III to accommodate only those students who were moving through the sequence in noncalculator sections. Full implementation began in the fall of 1994 for the 4,890 student enrollments in 134 classes each year of the above-mentioned courses.

Impetus for Change

Six years ago (1988), Clemson was delivering calculus in a very traditional way: lots of chalk-talk, no hint of technology, none of the other benchmarks that have come to be associated with the calculus reform movement (a focus on conceptual understandings, interactive learning, exploration and discovery, student projects, etc.). Like many others, our curiosity had been peaked by the report of the 1986 Tulane conference, *Toward a Lean and Lively Calculus,* and our eyes were opened wide by the 1987 National Academy of Sciences Symposium Calculus for a New Century. Our colleague John Kenelly had been a contributor to the Tulane conference and in 1987 managed to obtain a classroom set of HP-28C calculators on loan and taught Clemson's first calculator enhanced calculus. Indeed, it was largely the enthusiastic response of his students in this pilot course that persuaded us to take action. We were fortunate to obtain funding from the FIPSE program in 1988 and to secure additional HP-28 units on loan from HP. With some protracted arm twisting, we even got our local administration to purchase another 80 HP units to help begin the project.

Much of our early impetus came from the two conferences cited above, Lynn Steen's 1987 article "Who Still Does Math with Paper and Pencil" [2], and the new and emerging breed of hand-held technology represented by the HP-28S calculators. Although computing was commonplace with many of our faculty, it had yet to make an impact upon the way our undergraduates were taught and the way they learned. But it was the vision of high-level, symbol manipulating calculators—with features such as interactive graphics, menu-driven software, an enhanced operating system and effective memory management—in the hands of all students, that most captured our fancy. And Kenelly's students had responded with unprecedented enthusiasm to this vision.

Growth and Support

We began our move in the fall semester of 1988 with a core of six from among the fifty-three on the mathematical sciences faculty. Each one of the six was committed to the project and took on the major responsibility of developing a calculator enhanced version of an existing mainstream course. Nothing was available relative to the pedagogical use of high-level calculators, and microcomputer coursework was just starting to appear. Thus, our early work was largely a pioneering effort. During the first two years we taught prototype and pilot versions of our calculator enhanced courses with calculators loaned to students under signed loan agreements. We struggled with many issues, such as syllabi, texts, assessment and tests; how to get entering students started and proficient in their use of the calculators, and how to foster and maintain their active involvement with the course material; students' misconceptions about mathematics, the role of technology, and learning; and what were appropriate uses of the calculators to help improve learning. In July 1990 we published preliminary versions of brief course supplement manuals, and in the 1990–1991 year filled 33 calculator enhanced classes. In 1991–1992 the manuals were revised, and we jumped to 52 such classes, then to 56 in 1992–1993. We simply advertised that certain sections of our courses would be calculator enhanced and would require an HP-48 calculator; the students responded by filling up all sections to their maximum capacity of 40. During the period 1990–1992, we expanded from the original 6 faculty active in the project to 18 faculty and 17 senior graduate students, and we also began a series of weekly support seminars, informal gatherings of the instructors of the calculator enhanced classes. Seminars were held for each of single variable calculus, multivariable calculus and differential equations. These proved to be very successful and were the first seminars devoted to issues of undergraduate instruction to be held in our department in over 20 years.

Since 60% of Clemson's enrollments in single variable calculus are from students majoring in engineering, it was clear that we needed the active support of their faculty for our project to succeed. Engineers have always sought the highest level of technology for the practice of their profession, and professional engineers were the principal users of the generation of HP-41 hand-held units that preceded today's graphics marvels. We made a 90-minute presentation to the engineering dean, department heads, and curriculum committee in 1990 and followed up with visits to several to their departments. We even loaned some of our own faculty's HP-48 units to engineering faculty to help them become acquainted with its capabilities before they invested in their own units. The freshman engineering program was an active proponent of our calculator enhanced calculus after receiving enthusiastic reports from many of their students, and was instrumental in helping sell our early plan of voluntary enrollment to incoming freshmen students. In anticipation of our decision to require full-scale implementation in Calculus I during the 1993 fall semester, we made a second presentation to the dean, department heads, and curriculum committee in engineering in February 1993. This was followed by testimonials from senior engineering students and results from a recent questionnaire to new inductees of the engineering honorary society.

Chemistry also has found the HP-48's attractive for their majors, all of whom complete our four semester calculus/differential equations sequence. In fact, the chemistry department purchased 18 HP-48's to loan to any of their majors who in 1993–1994 may not have their own units; these will later be passed along to faculty within the department.

Student Perceptions

We have been unable to develop objective measures of the true learning that is actually taking place in our calculator enhanced courses. Content, approach, methodology, and student activity in our courses are all changing, and we have simply not had the professional expertise or resources to accurately analyze the extent to which these changes are being translated into more effective learning. Comparisons with traditional courses are not appropriate since expectations are substantially different, and grade distributions are of no help either. However, during the funding period of our project, 1988–1991, an external evaluator solicited student perceptions of the role of the calculators and the learning that is taking place in the new courses. These were obtained through a series of questionnaires completed by students at the end of their courses and in personal interviews with many of the students by the evaluator. A summary and discussion of the results appears in the article "Supercalculators in Undergraduate Mathematics" [1]. We have included as an Appendix some information from that summary. The data are from 1,523 students across 61 different classes of our calculator enhanced courses.

Lessons Learned

In looking back on the development and growth of Clemson's project over the last five years, there are some lessons for others who may also choose to embrace graphics calculators as part of their Preparing for a New Calculus. These are listed below, in no special order, and certainly with no claim of being thorough, only that they are lessons that we have often shared with others.

1. Commit to Change

The stimuli that spurred the Clemson effort in 1988 are still present; indeed, the calculators have become supercalculators, exceeding by far our most imaginative hopes. The reality is that impetus for change in calculus is now everywhere. From the strong NSF initiatives for funding in calculus reform, from the multitude of conferences, workshops and symposia held at local, regional and national levels, from the new generation of course materials that are being produced, and from the realization that many colleges and universities are involved in some way. Such impetus for change is impossible to ignore and hard to resist! The lesson here is that change requires commitment. And that commitment, whatever it may be, must be embraced with a passion.

2. Expect Faculty Resistance

Even with a strong impetus and a vision of the future, there is a high element of risk involved for faculty. Most faculty have had no training in techniques of effective teaching. They also grew up in an era before graphics calculators, and are often suspicious of their use. Some still view calculators as an inappropriate intellectual crutch, or as toys that have no place in "real mathematics." Asking them to invest in learning a new technology and the new approaches to teaching that it requires is, understandably, a threat to their time-honored way of doing business. Fear of change, lack of confidence, and visions of inadequacy are real human factors that must be faced.

3. Provide Support for Instructors

It is imperative that some sort of initial training in the technology and its use be given to instructors who will use calculators in teaching calculus for the first time. Often, only a one-day session is enough to get them started, because the calculator skills needed for first semester courses in calculus include only the following:

- produce "good" graphs of functions of one variable;
- use the graphics features appropriately (e.g., zoom, rescale, trace, find roots and extrema, etc);
- numerically evaluate functions of one or more variables;
- numerically evaluate Riemman sums and definite integrals.

For second semester courses, which typically deal with topics in integration, series, and special functions and curves (e.g., transcendentals, conics, polar and parameterized curves), the additional skills needed are:

- numerically investigate sequences and series;
- produce good graphs of the above mentioned special functions and curves.

Once a program has started, continuing support should be provided for the instructors during their first year, at least. Clemson's weekly support seminars are friendly forums for sharing ideas, experiences, examples of student work, what works well and what does not, etc.

4. Start Small ... but Think Big

If at all possible, begin with at least two class sections of the first course in calculus with two different instructors. That way, each instructor will have a supportive colleague with whom to interact as they proceed through the course for the first time. In addition to the struggles that were cited earlier, there are plenty of others to be faced.

5. Help Students to a Good Start

Effective implementation of hand-held technology requires that students develop reasonable skills in using their calculators in appropriate ways. Unfortunately, students do not come to us with these skills. Skill development is a process that unfolds naturally in the conduct of each course and should occupy little actual class time. Our experiences have shown us that any class time that is devoted to developing such skills is more than compensated for by the increased abilities of the students to engage the course material in effective ways. At Clemson, at the start of each fall semester we conduct two evening (7–9 P.M.) Calculator Help Sessions for incoming students. Typically 500–600 attend the first evening, and only half that many attend the second. Also, during the 1990–1991 year, we hired a senior undergraduate student to conduct walk-in calculator help sessions for calculus students two days each week.

6. Improve Incrementally

Preparing for a New Calculus does not necessarily mean that you must overhaul everything at once: course content, texts, methodology, and the introduction of technology. Granted, that may be the ideal way to go about it, but many schools across the country, especially those implementing calculators, are doing it incrementally. A common model is to implement calculators and modify the course content, topical coverage, classroom approach, and methodology with materials that supplement an existing text, preferably a text that is technology sensitive. In calculus, the primary benefit of technology is to encourage and promote multiple representations of limits, derivatives, and integrals by combining and comparing graphical, numerical, and algebraic points of view. And nowhere is this more effective and personalized than with supercalculators. These devices encourage a high level of personal interaction with the material on a daily basis and challenge our creativity by permitting an increased level of exploration. Supercalculators also promote and facilitate an increased level of mutual student/teacher interaction and serve to redirect the focus of learning from passive acceptance to active involvement. New, technology sensitive textbooks are now appearing, and most of them view supercalculators as tools in the service of learning mathematics. Your "new calculus" doesn't have to be perfect at the outset (it won't be) anymore than your "old calculus" was perfect when you first taught it. Remember: Change should be viewed as an ongoing process, not as an event.

7. Coalesce your Campus

Building and maintaining support from the major client users of calculus on the campus is absolutely essential, especially when there is a heavyweight user, such as engineering. This may not be easy, because most faculty in client disciplines also learned their calculus in a very traditional way. You will have to sell your ideas to them as an audience of nonmathematicians and show them some of the new dynamics of learning that you are working to achieve. They will also need time and opportunity to question and to try to connect the mathematics you are teaching to the mathematics that they do. But the rewards that come from their active support can be significant and far reaching, especially when you are planning to institionalize a program.

8. Seed your own Backyard

You can help your calculator program develop and grow by fostering similar programs in the local high schools and community colleges that send students to you. In South Carolina, for example, 30 of the best high schools now have full classroom sets of HP-48's, awarded free under a competitive grants program funded by NSF with corporate support from Hewlett Packard. Over 225 high schools across the country were the recipients of such awards in 1992–1993. By actively helping your local "feeder" institutions to develop their programs and showcase your own, you can help insure a steady stream of students into your program.

9. Sustain and Institutionalize Change

It is not sufficient to simply implement change; the change must be significant, productive, and long-lasting. It will not happen easily or quickly. The pathway is likely to be strewn with fragments of broken ideas, baited with false hopes, and lined with harsh realities. Sustaining and institutionalizing the desired changes in calculus requires patience and perseverance, great energy, and good leadership. The key to success will be the voluntary acceptance by the faculty of a shared responsibility for improving student learning.

References

1. LaTorre, Donald R., "Supercalculators in Undergraduate Mathematics," in *Impact of Calculators in Mathematics Instruction,* George W. Bright (Editor), University Press of America, 1993.

2. Steen, Lynn Arthur, "Who Still Does Math with Paper and Pencil?" *Chronicle of Higher Education,* October 14, 1987.

APPENDIX

Data from 1,523 students in 61 different calculator enhanced classes.

The graphics calculator helped me understand the material in the course.

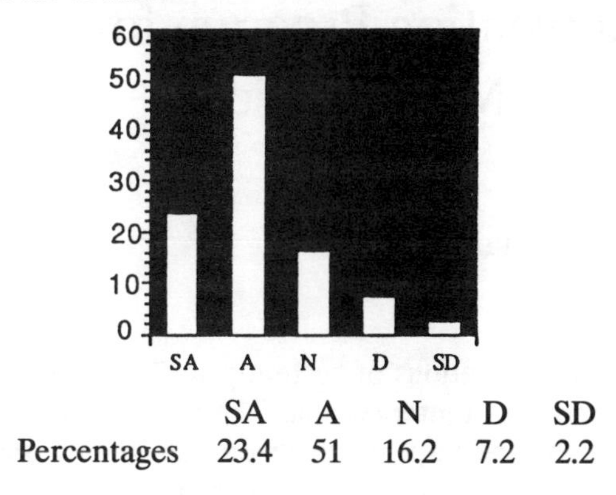

	SA	A	N	D	SD
Percentages	23.4	51	16.2	7.2	2.2

The graphics calculator allowed me to do more exploration and investigation in solving problems.

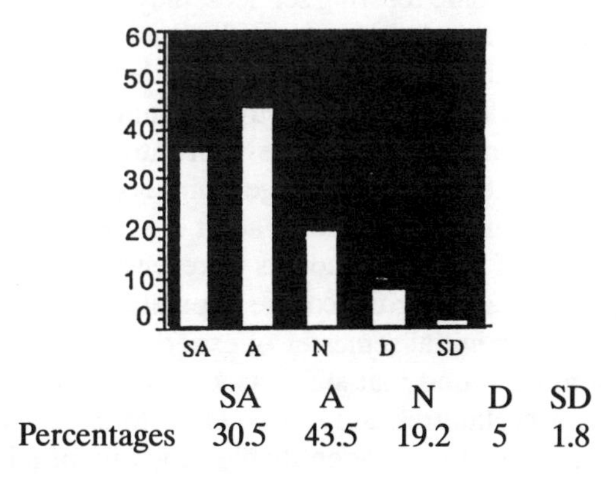

	SA	A	N	D	SD
Percentages	30.5	43.5	19.2	5	1.8

Learning the calculator was so difficult that it detracted from learning the material in the course.

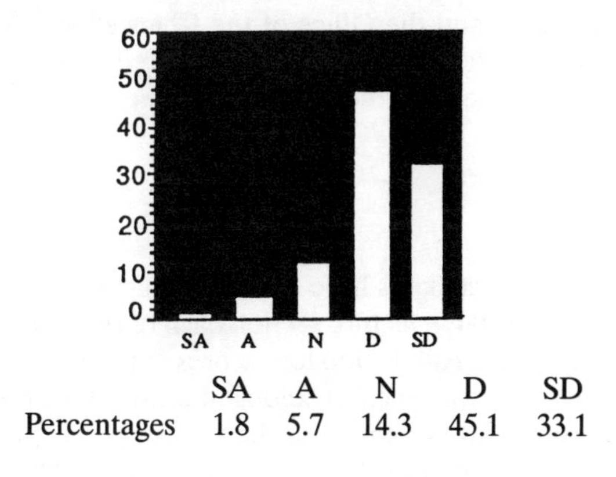

	SA	A	N	D	SD
Percentages	1.8	5.7	14.3	45.1	33.1

Time devoted to instruction in the use of the graphics calculator meant less material was covered in the course.

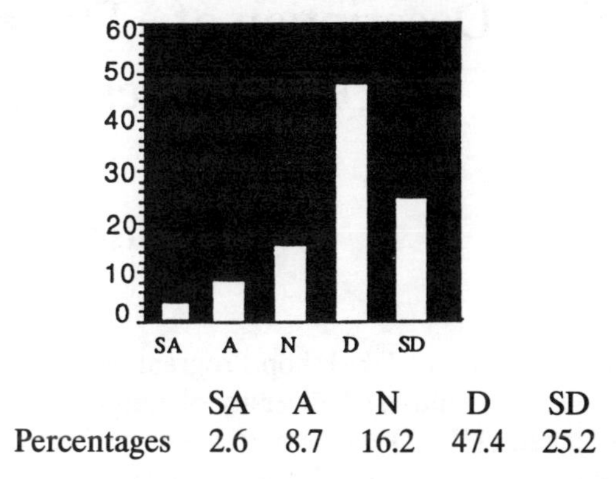

	SA	A	N	D	SD
Percentages	2.6	8.7	16.2	47.4	25.2

The graphics calculator helps me have better intuition about the material.

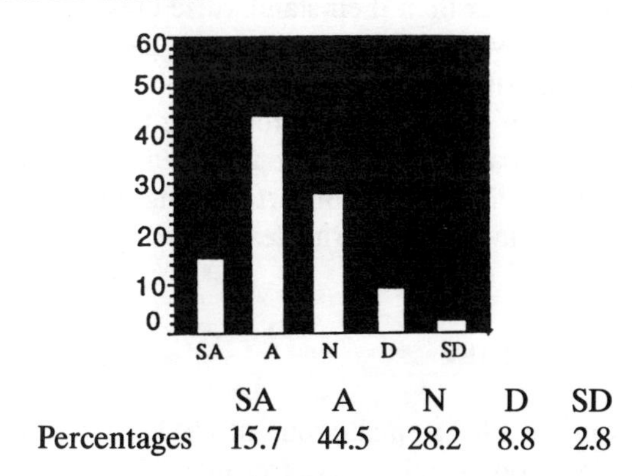

	SA	A	N	D	SD
Percentages	15.7	44.5	28.2	8.8	2.8

I would recommend that entering freshmen seek out courses using the graphics calculator.

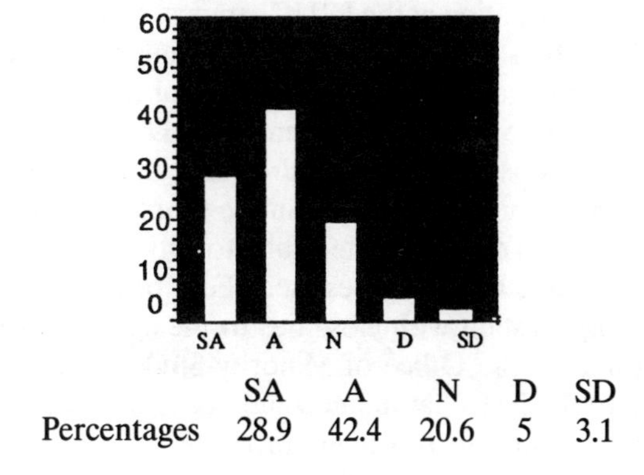

	SA	A	N	D	SD
Percentages	28.9	42.4	20.6	5	3.1

Legend: SA = strongly agree; A = agree; N = neutral; D = disagree; SD = strongly disagree

The Merit Workshop Program in Calculus at the University of Illinois at Urbana-Champaign: Description of a Successful Intervention Program for Underrepresented Groups in Mathematics

Paul McCreary
University of Illinois, Urbana-Champaign

Before the Merit Workshop Program began, minority students in calculus at University of Illinois at Urbana-Champaign (UIUC) did about as well as Treisman reported minority students did at UC-Berkeley prior to his Professional Development Program [1]. Their performance in introductory calculus courses was shockingly low, much lower than their standardized test scores predicted. The Merit Workshop Program has dramatically reversed the situation on our campus. On average its participants earn almost a full letter grade higher than the class. Participating cohorts have averaged as much as two grade points higher than nonparticipating cohorts of similar backgrounds and entering test scores.

Brief History

In Spring of 1987 a small group of UIUC staff attended a seminar conducted by Uri Treisman. During his talk Treisman gave a description of the environment facing minority students at Berkeley prior to the Professional Development Program. The description precisely fit the prevailing situation at the UIUC campus. Proportional to their numbers in the calculus courses, minority students earned less than half the expected number of A's & B's and over twice the expected number of D's and E's. Further, there was a much higher drop rate among these students than among all students in the course. At the seminar, Treisman described the efforts to initiate his program and indicated the resources used. Four professors and a teaching assistant with close ties to the existing support programs in the Office of Minority Student Affairs decided to see just what might come out of their combined efforts to replicate Treisman's program.

They recognized that the resources and structure to effect Treisman's dramatic results were available on the UIUC campus. They also knew from first hand experience the limitations of currently available, more traditional models of intervention. They had worked directly with intelligent students who were adequately prepared and who began the course with a positive, enthusiastic attitude. Even when such students had enrolled in the discussion section of a good and conscientious TA, and took advantage of some tutoring services, they often didn't excel and too often failed to pass the course. Of course, there were students who were only moderately well prepared, whose TAs were unable to communicate well, and who did not make connections with tutors. These students failed the course in large numbers. During any given semester it was not unusual for over 50% of the cohort group of minority students to receive a D or below compared to 25% for the course population as a whole.

During our initial semester of experimenting with the program, we found that students who attended our sessions had no failures and scored close to class average. This was a first for any identifiable subgroup of minority students on this campus. After these promising results, combined funds came from the Department of Mathematics, the Dean's Office in the College of Liberal Arts and Sciences, and the Office of the Chancellor. Over a two year period the program expanded to its present size of seventy students in the first two semesters of calculus.

Results

Workshop participants have scored the highest exam average in their large lecture section each of the past three years. This past Fall the top four scores out of all students in the large lecture section belonged to Merit Workshop students; almost two-thirds of the top ten percent of the grades were scored by workshop participants. The hour and final exams were all graded in common grading sessions including all teaching assistants and the lecturer to assure uniform grading standards. The participants of the Merit Workshop Program typically have standardized test scores slightly *below* that of all students from similar backgrounds and substantially below that of all students taking the calculus courses, making the exam results all the more remarkable.

Failure in calculus has been virtually eliminated among participants in the Merit Workshop Program: there are a few D's each semester and one E every other year. There are also virtually no dropouts among Workshop participants compared to over 30% among their cohort group. We have found that more minority students are choosing mathematics and mathematics education as majors. TAs and visitors to workshop sessions remark about how impressed they are by the quality of intellectual activity and the quality of human caring among the students.

Participants

The Merit Workshop Program is designed primarily for groups currently underrepresented in the science and mathematics based majors. This includes students from African-American and Hispanic backgrounds and graduates from small high schools. We invite entering first-year students from these groups who have chosen a math/science based major to participate in the program. Additionally, each year a small number of female students who are not from the other target groups participate in the program. Each year approximately 70 students choose to participate in the extra class hours and collaborative group activities of the program associated with first and second year calculus courses.

Structure

Participants in the program attend the same lectures, complete the same homework assignments and take all exams and quizzes together with other students in their large lecture section. The workshop participants spend an additional four hours per week in class sessions working collaboratively on problems significantly more difficult than the homework assignments. We neither grade nor correct this collaborative work, but use it to encourage student-to-student interaction and to allow immersion into interesting and open-ended calculus problems. Often, unless an exam is imminent, incomplete and even inaccurate ideas articulated by students in the group sessions are left uncorrected by the teaching assistant in charge. It has been our experience that most mistakes made by students as they worked in groups simply disappeared. The students are allowed and encouraged to explore and investigate their ideas even when these ideas are not well formed. They develop an enhanced intuitive understanding of the material and improve at persevering and making multiple attempts at difficult problems. Group collaborative activities are used as a device to encourage group interaction and to provide a context for intense intellectual activities.

Outline of academic program activities

The extended class hours allow time for students to grope without failure. Further, this process of struggling with the material often provides a natural context for filling in gaps left during their high school education. Without the need for any remedial activities, students teach each other necessary skills not mastered in high school. Students who already know the material serve as tutors to their peers, and they themselves develop more sophisticated and deeper understandings. A watchful TA is able to monitor the students' skills, and, when student-to-student tutoring proves insufficient, either provide on-the-spot tutoring or plan specific practice exercises for subsequent work sessions. The students come to see discussions about interesting mathematical problems and multiple attempts at difficult problems as the way mathematics is supposed to be. From these activities and interaction with their TAs and visiting faculty, the students develop a greater appreciation of mathematics as a discipline, a field of study, and a profession. TAs in the workshop program develop a keener insight into what is actually going on in their students' minds, a valuable tool for any instructor to have.

Description of a Workshop Session

Each of the workshop sessions is two hours in length. An observer of any portion of the session would immediately note the informal nature of the interaction between the students and their teaching assistant and among the students themselves. Tables of various shapes are arranged about the classroom. Each has enough work space for about eight students. Chalkboards cover three walls of the room. Pictures of all the participants are mounted above the chalkboards, grouped by class sections.

As students arrive, they arrange themselves around the room. The groupings at tables are either by the students own choice or by the teaching assistant's assignment, depending on what the teaching assistant has planned. In any case, once work has begun, students move freely from table to table, seeking assistance for their own questions or responding to a request from a fellow student across the room. Often students move from working at the tables to the chalkboards.

"Rhythms" Within a Workshop Session

Each Workshop session has a definite internal rhythm and structure. The first 20–30 minutes is a warm-up period. During this time students say hello to each other and to the material in various ways. They may exchange anecdotes about what happened since they last saw each other, joke together, perhaps even talk about a homework problem they previously worked on together. Likewise, they look at the material and consider it again seriously after an extended period apart from it.

We have noticed that it seems to make very little difference which students spend the initial warm-up time working on mathematics and which spend the time chatting about unrelated topics. As long as almost everyone is actively working on either reestablishing familiarity with the mathematics or their familiarity with each other, and as long as sufficiently many (about half) are engaged in the mathematical work, then *everyone* seems to make substantial progress during the intense, middle work period. It seems that those who have been considering the mathematics help the others to catch up with them. The personal connections reestablished during the warm-up period seem to make dissemination of new mathematical revelations smoother and more rapid. In this way everyone is able to participate fully in the mid-session work period, which lasts well over an hour.

When a problem proves particularly baffling, a group may gather around the most active movers, especially when other students sense that significant progress is being made. It sometimes develops that a majority of the class congregates at one table amidst a flurry of activity. A student looking over another's shoulder for a moment might turn excitedly back to an earlier collaborator to indicate that their group *had* been on the correct track. Another trio of students might move to a chalkboard to try another idea that occured to them. This type of intensive activity usually occurs only after students have been working on a particular problem for an extended period of time, once they are familiar with the question and are clear that the problem poses a serious challenge.

During the last twenty to thirty minutes of the two-hour session, most of the students are becoming fatigued and have a much reduced capacity to concentrate on the material. On the other hand, their familiarity with the material is at its peak. It is during this period that a TA can take advantage of observations made throughout the work period. The TA can find whether a mistake was a mere slip of the tongue on the student's part or an indication of a deeper misunderstanding. A student who has clearly understood the worksheet problems can be lead further into underlying concepts in a brief one-on-one discussion with the TA, perhaps with a small group of students as interested spectators. Such engagements with the students can be among the most exhilarating for any TA. The students' familiarity with the particular ideas is at its greatest; the opportunity to synthesize this material may not arise again for several years, when the student sees it again in advanced calculus courses. It is a unique opportunity to introduce some of the beauty and mystery of mathematics which must be postponed in most contexts because of the high investment required to provide the background before students are able to understand the arguments and discussion.

Description of Activities During a Semester

In our program, for approximately the first third of the semester the focus of the teaching assistants' energy is around issues other than the course content. This is not to say the teaching assistants ignore the topics being covered in the course. However, the most important task at this point is to facilitate connections among students in the workshop section. There is one overwhelming reason for this focus, which has to do with why students studying in intensive collaborative groups are so successful compared to students who attempt to learn the course solo. During the last third of the semester, a period notorious for holding the most difficult and complicated applications of the course and a period during which students have the least time for reflection or assimilation, students connected to a group who have learned to discuss course related issues and ideas with each other are in a position of enormous advantage. It has been our experience that during this late period in the semester a teaching assistant can effectively teach the entire workshop class by showing a select two or three students a new idea and then watch the idea spread throughout the entire class. Once one student in the group of 20–30 students learns a new idea, all members of the section learn it at an almost unbelievably rapid rate. The abilities of the group as a whole to inform all of its members of important, intricate facts and methods reaches its peak at precisely the moment when it is most valuable for its individual members.

Perhaps this process of establishing collaborative work groups will occur more or less spontaneously when enough of the student body enters the university with a background of effective collaborative work in high school. However, this is not now the case, and a concerted effort is necessary to help students establish connections.

During the middle third of the semester the content of the course is emphasized along with the importance of student-to-student interaction inside and outside the workshop, during homework and review study times. There is a tendency during this period for students to remain within relatively small sub-cliques within the section, with those whom they met first and with whom they feel most comfortable. Considerable efforts are expended to assure that there is substantial communications among these sub-groups. For the most part, students sincerely wish to participate in this increased interaction among their peers. The teaching assistant's insistence on the interactive process and on the group activities legitimizes the interaction in the students' minds and affords opportunities for the increased interaction. So, despite the inevitable reluctance to further mix among their fellow students, the teaching assistant always finds widespread cooperation with this process.

During the final third of the semester, several issues related to study habits become the primary focus of the teaching assistant's efforts. Other concerns are emphasized, including staying healthy, staying caught up in all courses, especially not studying a single subject prior to that subject's major exam to the exclusion of all others. Throughout the semester, problems are selected to assure a process of review and preview of the course material.

Recruiting Activities

On the UIUC campus, approximately 350 students from minority backgrounds and 75 from small high schools are invited to apply to participate in the program. Eighty-five to ninety students respond to the invitation, and 70 become active participants in the Fall semester program. The majority of participants each year are entering first-year students who have already selected mathematics or science-based majors. They are notified of the program during the spring of their senior year in high school and are met on campus during their summer orientation and advance enrollment activities. During this past year a few participants were identified and recruited through recommendations from instructors in prerequisite courses. More than a few were recruited directly by current or former participants from among their peers.

Teaching Assistants: Their Role in the Workshop

Paradoxically, it is their most well-developed skills that TAs must suppress in order to be effective in the collaborative classroom. TAs who choose to work in the Merit Workshop Program are often the most experienced in the department. Because of the opportunities in traditional settings, the TAs have well-developed skills in planning and delivering coherent lectures, tutoring small groups during office hours, and providing thorough answers to students' questions. In collaborative classrooms, instructors must refrain from answering questions. Instead, they must engage an ever increasing number of students and help *them* to address the questions asked by their fellow students. Instead of correcting mistakes voiced by students, TAs must bite their own tongues and remain alert for opportunities to bring students with better understandings into the conversation.

Good TAs can have a deep and significant effect on *five* times the number of students in a class structured around collaborative group work compared their effectiveness in more traditional settings. This is one of the greatest attractions for TAs who choose to work as workshop section leaders on this campus. On the other hand, adapting their teaching styles to effective workshop techniques is not easy. An experienced TA in the program has noted that he and other TAs in the program accomplish most of the more traditional teaching activities as they compose the daily worksheets of challenging problems. In this activity they "teach a session before it happens." The logical steps brought out in a well-organized lecture, the exercises, examples, attention brought to the prerequisites of new material, and all the components of an explanation are injected into the material in a good worksheet. Of course, in reality the students take over the material, rearrange the order, even ignore the points most important in the minds of the TAs. An effective TA allows this to happen. The efforts to arrange the material in a logical order are not wasted. They give the TA a better, more thorough understanding of the material's underlying structure. The TA can then see actions by the students, their revelations and misconceptions, in a broader context. The TAs place themselves in position to understand more of what the students will be trying to do and they will be in a better position to help the students along their way. They do this rather than attempt to steer the students along a lesson plan's predetermined, narrow corridor. The students, wheeling through their own agenda, learn many things, even review prerequisite material and construct connections that the TA never anticipated in the lesson plans.

TAs should certainly remain aware that once they say "no," some listeners immediately believe that *everything* said before was wrong. Further, student conversations will always be riddled with mistakes. Some of the mistakes are mere slips of the tongue while others are not at all central to the discussion. It's improper to disrupt an arguing student's momentum by insisting on verbal consistency.

In our experience, a voiced mistake by a student in a workshop is almost assured not to cause long lasting harm to other students who are listening. Even if the student is tutoring another student and is "teaching" misinformation, no permanent damage will occur. In many, many cases people simply don't listen effectively, so it is often OK to allow students to "teach" each other mistakes. If a student is nodding enthusiastically in the midst of an incorrect lesson, quite often the affirmation means, "Yes, I recognize and finally am understanding the terms being used." The student is not understanding and often not even listening to how the concepts connect with each other. The misinformation is most often at a higher level of understanding than the listening students are yet able to comprehend. In a sense the mistake "goes over their heads" and is lost. What they do gain from the interchange is practice listening to and thinking about the terms or ideas in a rudimentary way. Even if the explanation were perfect in every detail, they would probably not carry away anything close to a thorough understanding. Likewise, they do not carry away a thorough grasp of most misinformation. As they continue to work and discuss, the mistake may never again be repeated. If the verbalized mistake does spring from a misconception, well-developed lines of communication among the students will ensure that once the correct understanding is firmly established within any subgroup of the students, it will spread to all others in the class. The understandings that develop in this way are seen to be lasting. Some interesting, even amusing situations have developed about shared misconceptions in the short run, but we have yet to witness a similar dissemination of misinformation that was lasting.

Another reason for a TA wishing to interject comments into a student conversation is often the perception by the TA that the students will surely understand what's going on so much more if only they are told or led to see one particular fact. An alternative to presenting this fact to the student group is to realize that a *student* may be on the verge of noticing this fact or idea. A TA should watch for a student to come to this revelation and be ready to help it spread among the other students.

Certainly a TA shouldn't remain mute, rather, he or she must learn how to *not* give the mathematical content while at the same time strengthening student enthusiasm for pursuing questions. Thus, TAs must cultivate the important skill and art of *not* answering questions while at the same time keeping the conversation going in fruitful pursuit of the problems.

Final Comment

The participants are not weak mathematics students. There is a common misconception that students in the Merit Workshop Program are weak mathematics students. However, as is the case with the general student body on the UIUC campus, well over half the Merit Workshop participants graduated in the top ten per cent of their high school class. They have good reason to believe that they are exceptional students. As Treisman noted at Berkeley, a determining factor for failure in calculus among students from our target groups is not poor preparation nor is it low skill levels. What the causes are for such shockingly high failure rates among minority students remains a largely unanswered question. However, it is clear that in the context of an intensely interactive group of peers with an enriched diet of challenging mathematical problems, these students not only survive, they thrive.

Reference

1. Treisman, P. Uri, "Proceeding of the July 6–8, 1988 Workshop, Mathematicians and Education Reform," Providence, RI: Conference Board of the Mathematical Sciences, *Issues in Mathematics Education*, **1**, 33–46, 1990.

A Gateway Testing Program at the University of Michigan

Robert E. Megginson
University of Michigan at Ann Arbor

The Department of Mathematics of the University of Michigan at Ann Arbor is implementing a reformed calculus project, based on the Harvard Consortium calculus text [3], through an effort directed by Morton Brown. Like most reformed calculus project implementations, ours deemphasizes routine manipulation and drill in favor of far stronger emphasis on insight and understanding. Though this redirected emphasis should lead to better calculus courses, it does cause some lingering concern that the students will not exercise and learn whatever routine manipulative skills really are necessary for a mastery of calculus.

As became clear during the course of the conference whose proceedings are published in this volume, there is not yet any universal agreement on what manipulative skills are actually important for students taking a reformed calculus course. One example cited in the draft report of the Content Workshop of the conference is the following: Should students be able to add $1/(x-1)$ and $1/(x+2)$ by hand and more complicated fractions using a symbolic manipulator, or should they perform all such manipulations by machine but be able to check if a result is reasonable? Answers to such questions may become somewhat clearer as the calculus reform movement progresses, especially now that some of the reform emphasis is shifting from calculus itself to the precalculus courses specifically designed to prepare students for a reformed calculus course.

Once a core of manipulative skills necessary for calculus is identified, the next question is how to teach those skills and assure that they have been learned. Of course, the traditional method is through a sizable collection of drill questions assigned as homework, followed by the testing of those skills on portions of standard in-class examinations. The main difficulty with this approach is that if enough such questions are assigned to assure that these skills are mastered by almost all of the students, then most homework questions tend to be these drill exercises, from which many students quite reasonably conclude that the main point of the course is to teach such routine skills. The large amount of time spent testing those skills on in-class examinations reinforces that conclusion. Another difficulty is that students can fail to master skills that are actually quite important for calculus, yet still receive an adequate grade in a precalculus or calculus course by learning other material well enough to offset the loss of points on tests due to the imperfect mastery of the manipulative skills.

An alternative way to assure that students acquire specific manipulative skills is through gateway testing. In this paper, the term *gateway test* will refer to a proficiency test of manipulative mathematical skills that must be passed as part of the course requirements if the student is not to incur some fairly substantial penalty. The name refers to the fact that the test is a gateway through which a student must pass to complete the course satisfactorily (and, if not passed, is a barrier to such satisfactory completion, for which reason the more ominous term *barrier test* is occasionally used for such an examination). Gateway tests are used in a number of institutions; for example, the United States Military Academy has a fairly extensive gateway testing program to assure that their students have the precalculus skills necessary for success in their mathematics, science, and engineering programs [1, 2].

In many implementations of gateway testing, the test, or variations of it, may be repeated without penalty as long as it is eventually passed. In this paper, gateway tests will be assumed to have this feature. For such a test, it is not necessary to drill students over the skills to be tested until almost all have enough mastery to pass, thereby boring those who caught on quickly. The test can be administered after the students have been given a much smaller number of exercises to work out, and then those who do not pass can study the material further, work out additional exercises, and perhaps receive some individual attention from the teacher or a tutor before attempting the test again. If the skills to be mastered should have been acquired in a previous course, then the test can be administered without directly covering the material in the current course, and remedial efforts can be concentrated on those who do not pass.

A logistical problem can arise after the first administration of a gateway test in class. If further attempts at the test are to be given in class and some of the students have already passed it, those students will need to be given something to keep them occupied while the others retry the test. On the other hand, if retries are given by the teacher outside class, scheduling problems can arise for both the teacher and the students. These difficulties can be overcome by having the gateway test available on demand at a testing center. This also prevents valuable class time from being taken up by the gateway testing program.

During the winter, 1993 semester at the University of Michigan, we conducted an experimental gateway testing program in our standard precalculus course in anticipation of continuing such testing in the implementation of a reformed precalculus course in the fall of 1993. We also experimented with gateway testing in our second semester reformed calculus course. For the precalculus course, the program consisted of five short examinations over routine algebraic and trigonometric skills, for each of which the students were allowed ten to fifteen minutes. All of the material to be tested was covered in the course. For the second semester calculus course, one thirty-minute examination was given over antidifferentiation, material that had been covered in the first semester calculus course. The rest of this paper will deal with the more extensive precalculus gateway testing program, but many of the observations made about that program also apply to the other.

The topics covered by the five precalculus gateway tests are: (1) elementary algebraic skills such as the addition of rational expressions; (2) further elementary skills such as applications of laws of exponents; (3) the solution of linear and quadratic equations; (4) the solution of more complicated types of equations; and (5) trigonometry and laws of logarithms. Test 4 had four questions on it, while each of the others had five. On each test, the only passing grade was a perfect score, and "small" errors were not forgiven. A typical question on logarithms from Test 5 would be similar to the following one:

$$\text{Completely expand and simplify } \ln\left(\frac{x^7\sqrt{e}}{(x+y)^{19}}\right).$$

For this problem, "completely expand and simplify" means just what one would think it should and was a well-defined term for the students. Such a question might be inappropriately manipulative for a reformed precalculus course, but it should be kept in mind that this question was for a standard precalculus course taught from a traditional text. Though easy-to-grade multiple choice tests were an attractive option, it was decided that students should be allowed to work out and present their own answers rather than selecting them from prepared lists.

All gateway tests were given in our Mathematics Laboratory (MathLab), a facility whose main purpose is to be a free no-appointment tutoring service. A student wishing to take a gateway test would request the particular test from the student manager of the MathLab, who would verify the student's identity from the student's university identification card before administering the test. After the student finished, the test would be graded immediately by a tutor, who would go over any errors with the student and offer tutoring to the student on whatever test material the student did not understand well. To encourage students to study the material between tries, each student was allowed only one attempt per day at each test, with the lone exception being that if only one question were missed, then the student would be allowed to retake the test immediately.

Since the students could retry the tests, it was necessary that each test be available in different versions. Rather than writing a large number of different versions of each test and incurring the record-keeping required to monitor which students had taken which versions of each test, we elected to use computer-generated tests. For each type of question to be asked on a test, a group of ten similar questions was written in TEX-compatible format using a word processing program that could output an ASCII file. Thus, for a five-question test, five groups of ten questions were created. These questions were then imported into a Lotus 1-2-3 spreadsheet containing macro code that would construct a test by selecting one question at random from each group of ten and writing out an ASCII file ready to be processed by TEX. The Lotus 1-2-3/TEX combination was selected because of the wide variety of computer platforms on which both of these products are available. In fact, this became important to us one day when the MathLab's IBM-compatible computer broke down, and it was necessary to run off copies of the test in a hurry using a SPARCstation connected to our department's Ethernet LAN. Directions for constructing the skeleton spreadsheet, along with operating instructions, are available from the author.

The incentive for a student to pass the gateway tests was that the student's course grade would be lowered by one fraction of a letter grade (for example, from B+ to B or from C– to D+ in Michigan's +/– grading scheme) for each gateway test not passed by the end of the semester. This penalty was considered to be severe enough to get the students to put forth a good effort to pass the tests, but not implausibly draconian. In fact, 131 of the 153 students completing the course passed all five tests, and of the 22 students who did not, ten passed all tests but one. The students clearly took the tests seriously.

Questions were included on the common midterm and final examinations specifically to test the material covered by the gateway examinations. In particular, one set of questions on the midterm and final examinations, which will be called Question Set A, was designed to test the same skills as Gateway Tests 1 and 2, while another set, Question Set B, was designed to test the same material as Gateway Tests 3 and 4. The difficulty level of Question Set A on the midterm examination was about the same as that of Set A on the final examination, while Set B on the midterm examination was somewhat easier than Set B on the final examination. With only a few exceptions, the questions were graded as harshly as possible, with little partial credit given.

There was no control group of students who were not required to take the gateway tests, so it is not possible to establish scientifically what gain in scores on Question Sets A and B can be ascribed to the gateway tests. However, there were imperfect, self-selected control groups composed of students who had not attempted gateway tests before the midterm examination but had passed them by the time of the final examination. Let Group U12 consist of those who had attempted neither Test 1 nor Test 2 before the midterm but had passed both by the time of the final examination, and let Group P12 be those who had already passed both tests before the midterm examination. Let Groups U34 and P34 be the corresponding groups for Tests 3 and 4. The sizes of Groups U12, P12, U34, and P34 were 17, 46, 53, and 14, respectively. The following table gives the average scores of the members of Groups U12 and P12 on Question Set A of the midterm and final examinations.

	Average Score on Midterm Exam Question Set A	Average Score on Final Exam Question Set A
Group U12	38.7%	77.3%
Group P12	73.9%	79.5%

It can be argued, probably correctly, that Group U12 consisted of less motivated students than Group P12, since the members of the first group had not even attempted either Test 1 or Test 2 by the midterm while the members of the second had already passed both, and that this difference in motivation accounts for much of the difference in the scores on the corresponding material of the midterm examination. In that case, it is even more impressive that the average skill level of Group U12 on this material almost matched that of Group P12 on the final examination after both groups had passed the two gateway tests. As can be seen from the following table, the general trend was the same for Groups U34 and P34 on Question Set B of the two course examinations.

	Average Score on Midterm Exam Question Set B	Average Score on Final Exam Question Set B
Group U34	66.0%	82.8%
Group P34	88.2%	79.6%

Again, there was a fairly large difference in skill levels at midterm time, but an almost insignificant one at the time of the final examination when the members of both groups had passed the two gateway tests (and in fact the group who had not even attempted the tests by midterm actually outperformed the other group slightly, perhaps because its members had mastered the material more recently). It was not possible to conduct an equivalent evaluation for Gateway Test 5, since it had not even been offered by the time of the midterm examination. However, a question set on the final examination did test the skills needed for that gateway test. The average score on that question set was 74.1%, which we thought was acceptable considering the difficulty of the questions.

From the results of the preceding paragraph, it seems that the gateway tests did what they were designed to do, which was to force the students to master the manipulative skills being tested. By using such gateway testing in our reformed precalculus program, we should be able to avoid stressing such skills in class or directly testing them on any in-class tests or the common midterm and final examinations, which will give us much more time to deal with the real conceptual material of the course.

Several mistakes made in our implementation of gateway testing should be mentioned. Gateway Tests 1 and 2 were originally presented together as one ten-question test. From the frustration level of the students, it quickly became apparent that a twenty-minute ten-question test is too long if the students are not allowed to make any mistakes on it. Rather than lower our expectations by allowing the students to miss a question or two, we decided to split the test into two ten-minute five-question tests. We feel that we accomplished the same goal that we would have with the original test, and the students were much happier.

Another mistake was allowing the students the entire semester to pass each test. The predictable result was procrastination by many students, and an almost overwhelming rush in the MathLab the last week of the semester as many finally got around to finishing up the tests. In the future, there will be a much shorter time, perhaps two weeks, between the initial and final offering of each test.

Still another mistake was not allowing the students to keep the completed examination sheets to study. It was felt that with only ten questions in each group from which an individual question on an examination was taken, allowing the students to keep the examinations would quickly result in someone's compiling the relatively short list of all possible test questions. While this probably would happen, one solution would be to increase the size of each group of questions to, say, thirty; a list of all 150 questions that could appear on a five-question test would not be of much use to a student unless the student took the time to master all of them, in which case the test would probably have achieved its purpose anyway.

The resources needed to run a gateway testing program should not be underestimated. Our program was fairly labor-intensive, with all tests graded by hand. Since our tutors were supposed to discuss the students' errors with them as the tests were graded, it is difficult to see how the grading could be automated while retaining this feature. However, if it were decided to forgo this aspect of the program, then it would be possible to conduct the gateway tests at computer terminals, perhaps with the further concession that the tests be multiple choice, allowing the mechanization of the testing and grading process and the recording of scores. Without such automation, a program such as ours requires about 42 grader/tutor-hours for each gateway test given in a 100-student course, assuming that the students need an average of about 4.2 tries to pass and that the grader/tutor requires about six minutes to grade and discuss each attempt, as was our experience. Additional resources are needed for test administration and record-keeping.

On the whole, we are quite happy with the results of the testing program. By isolating the manipulative skills we wish the students to master on tests that the students are strongly motivated to pass, we are assuring that the students will not avoid learning essential manipulative skills that seem difficult to them and attempt to compensate through thorough mastery of easier material. Since the students have to retake each test until they get it right, they are forced to devise strategies to fill in gaps in their knowledge, which is an essential skill for students taking a reformed calculus course in which lecturing is deemphasized. Furthermore, by forcing the students to develop these strategies and removing the testing program from their classrooms, it is possible to use class time to concentrate on the conceptual material that should be the heart of a mathematics course. For all of these reasons, we look forward to making gateway testing an integral part of our precalculus and calculus reform programs.

The author would like to thank Beverly Black of the University of Michigan's Center for Research on Learning and Teaching for her valuable suggestions on evaluating our gateway testing program.

References

1. Arney, David C., Department of Mathematical Sciences of the United States Military Academy, private correspondence and transparencies from a presentation at the Winter, 1993 Joint Mathematics Meetings in San Antonio, TX.

2. Giordano, Frank R., *Core Mathematics at USMA*, Department of Mathematical Sciences of the United States Military Academy, West Point, NY, 1992.

3. Hughes-Hallett, Deborah, Andrew M. Gleason, et al., *Calculus*, Wiley, New York, preliminary ed., 1992.

Faculty Development for Using Technology in Teaching Calculus

Don Small
U. S. Military Academy

Fundamental changes in undergraduate mathematics instruction are sweeping across the country. These changes are driven by the increased payoff in student learning from the use of technology in teaching and from changes in pedagogy that emphasize student involvement. They are reforming calculus courses at an increasing number of institutions from high schools with Advanced Placement programs to large universities. However, in many instances within these institutions the reformed course is the product of a few people (1–3) and its offering is restricted to one or two sections. The purpose of this paper is to advocate a faculty development program leading to department-wide acceptance and implementation of a reformed calculus program, particularly one that uses technology, such as computer algebra systems (CASs) or graphing calculators.

The proposed development program is described in three steps. The first step is to develop awareness of local conditions and of the underlying characteristics of the national calculus reform movement. The second step is to develop a syllabus for a reformed calculus program that includes the use of technology. The third step, which is by far the most difficult, is to implement the reformed calculus program. Although the beginnings of these steps are linearly ordered, the steps overlap. The need for enhancing awareness as well as the need for monitoring the calculus program is never finished. Many of my comments will be based on my experiences as director of the National Science Foundation's program of national and regional CAS Workshops over the past five years. Although the focus of this paper is on creating an environment that is conducive to accepting and implementing fundamental changes in content and pedagogy in the teaching of calculus, I expect my comments to apply to other mathematics and science courses as well.

Now for some elaboration on Steps 1–3:

Step 1: Awareness

Understanding the contexts of a problem is the important first step in problem solving. For curricular changes, the contexts form a nested set. From the inside out, these are: student, instructor, department, institution, and, finally, nation. At each level we need to ask: What are the goals and objectives of our calculus program? Does our program address these goals and satisfy their objectives? Are we satisfied with our present calculus program? If the answer is "No," then how should it be changed? (In this paper, I assume the answer is "No.") How can technology be used to enhance student learning?

The "satisfaction" question has been answered in the negative for both the national and instructor settings. International assessments and studies [6, 7, 8, 9, 10] show the undergraduate mathematics program in general, and calculus in particular, is not adequate to support the business and industrial requirements of our country. As a result, the National Research Council and all of the professional mathematical associations support the calculus reform movement. The use of technology, particularly CASs, for exploration as well as computation is strongly recommended by these groups. At the instructor level, the answer given to the "satisfaction" question at the Tulane conference [2] in 1986 was clearly "No." This answer was underscored the following year at the Washington conference [14]. These conferences, bolstered by reports from the National Research Council, launched the calculus reform movement. The "satisfaction" question at the institutional and departmental settings has often been ignored for a variety of reasons, mostly political.

A primary task for those interested in introducing the use of technology into the calculus program is to create an awareness of the calculus reform issues within both the department and institutions. Doing so is a communication problem. Some strategies for doing this include:

- Obtain and circulate reports published by the National Research Council and professional mathematical associations (see the list of references) to appropriate administrators and colleagues including those in client disciplines.

- Post articles on bulletin boards. (Highlighting appropriate passages before circulating or posting helps focus attention on pertinent issues.)

- Involve students in building awareness by holding discussions with student groups such as MAA Student Chapters, math clubs, teacher education students, etc.
- Invite speakers to talk on calculus reform issues as part of a departmental or divisional colloquium program. Remember to invite administrators and colleagues from client disciplines.
- Hold one-on-one discussions with colleagues. This is probably the most effective technique. In particular, the well-respected, tenured members of the department need to be involved in these discussions.

An essential objective of the awareness building program is the development of a departmental statement of goals for the calculus program. The absence of such a statement prevents true accountability of the program. Unfortunately, few departments have a statement of goals for their calculus program. Approximately two thirds of the participants at the National Science Foundation sponsored CAS workshops indicated that neither they nor their departments have such statements. The old adage: "If you don't know where you want to go, then there is no problem" clearly applies.

Step 2: Develop a Syllabus for a Reformed Calculus Program

As mentioned above, the first objective is to develop a departmental consensus on a statement of goals for the calculus program. The statement must be explicit enough so that individual goals can be translated into specific enabling objectives that can be measured over both the short term (within the courses) and the long term (in subsequent courses). In addition, each goal and objective should be accompanied by a short description of applicable content, pedagogy, and technological tools. I include here, as an example, the set of goals that I distribute to my students during their first class.

Helping students to:

- learn how to learn mathematics;
- learn how to analyze functions;
- develop an inquisitive and exploratory approach to learning mathematics;
- become engaged in doing mathematics; and
- develop an understanding and appreciation of the 3 major concepts in calculus:
  - closeness (approximation and error bound),
  - differentiation (rate of change),
  - integration (partitioning, approximating, summing).

These goals must be translated into specific objectives. For example, I include the following under the fourth goal listed above.

Exploratory class activities and assignments Example: Develop a recursive formula for $B_n = \int x^n \log(x)dx$ following the paradigm:

(i) Generate data, i.e., solve for B_n for $n = 1, 2, 3, 4, 5$. Express your answers in terms of B_n for smaller values of n.

(ii) Conjecture a formula from the results of part (i). (Generate more data as needed.)

(iii) Check your conjecture by evaluating B_n for a value of n not used in your data collection.

(iv) Prove (or disprove) that your conjecture is a theorem.

Student construction of examples or counterexamples Example: Construct an example of a bounded function whose first derivative is unbounded or show that such a function cannot exist.

Individual and group projects Examples: Determine the number of zeros at the end of $n!$, develop a "comfort function" for stairways, approximate the surface area of a pond, and/or approximate the length of a trail.

Student presentations of course topics to the class Examples: volumes of revolution, L'Hôpital's Rule, and/or Ratio Test.

Developing a statement of goals is a modeling problem that will require several iterations to reach a satisfactory and doable statement. Involving colleagues in client disciplines, administrators, and students in developing a statement is extremely important in terms of acquiring institutional support. Once obtained, the statement should be used to guide the development of a syllabus for a reformed calculus program and also to provide a basis for evaluating the success of the program. The statement should be shared and discussed with colleagues in other disciplines, administrators, admissions people, computer services people, and the athletic department. The purpose is to gain widespread support both inside and outside the mathematics department. In particular, it is important to have support from the appropriate administration officials, as the statement may serve as a basis for requests for equipment, released time, or additional staff.

Using the statement of goals to develop a syllabus for a calculus program places the focus on what topics need to be included rather than on the "unanswerable" question of what topics should be omitted. An extremely important constraint to impose on the syllabus is to restrict the "core" content to what can be covered in, say, 60% of the available time. (It can be done.) This allows for several things to happen, such as:

- providing each instructor with time to enrich the courses by augmenting the treatment in his or her unique way;
- providing "negotiating room" when deciding on the core content;
- assuring time for covering the core content and thus easing the transition to the following course;
- creating good will among the faculty.

Providing each instructor with the time opportunity to "individualize" his or her course may be the most important ingredient in developing an interesting, rich, and vibrant calculus program.

Step 3: Implementation (4 Prongs)

The goal is the implementation of a reformed calculus program throughout the department based on the statement of goals. This will take time as it involves changing attitudes and probably reordering priorities. Planning on a three to five-year implementation program starting with an evolutionary, multi-pronged approach holds a greater promise for achieving the final objective than pushing for a one or two year accomplishment. Four important prongs to cultivate for this process are: student prong—developing student interest in using technology for doing mathematics outside the classroom; colleague prong—assisting colleagues in developing expertise in using technology for doing mathematics; course prong—experimenting with the use of a technology as a teaching tool in a calculus course; resource prong—procuring resources.

1. Student Prong

Students are important allies in the campaign to bring technology into the classroom, and their assistance should be carefully cultivated. I try to engage first year students in doing mathematics outside of a formal course. Technology provides a nice "hook" for doing this. An effective tactic for "setting the hook" is to have a couple of interesting exploratory problems ready to suggest when the opportunity presents itself. For example, how many petals does the polar plot of $r(t) = \sin(3t)$, $0 \leq t \leq 2\pi$ have? How about $r(t) = \sin(4t)$? How about $r(t) = \sin(nt)$? Another fun example is to use a CAS to expand $n!$ for a few values of n, observe that (for $n > 4$) the expansions always end in a string of zeros (Why?), and then ask: How many zeros are at the end of $n!$? To create the "opportunity," I first identify students (two is the best number) whom I want to "catch" and then "accidentally" meet them in the hall and start a conversation. The talk will eventually lead to my saying something like, "Say, I found a neat problem that I want to show you." We soon move into my office, lab, or classroom where there is privacy and a chalkboard on which the problem can be sketched. In a few minutes we are at the computer, preferably with the student at the keyboard. I stay with the students until I am confident that they understand the problem, have developed an approach to it, and have identified two or three related questions. We then agree on a time to meet in the next day or two. (The first problem should be nontrivial, but doable in less than an hour.) When the students return, we discuss what they have done, their thinking processes, what worked, what didn't work and why it didn't, as well as some related questions. If the students have obtained a satisfactory solution, I will start them on a related problem and set up another meeting. For some students this activity may go on for months, for others it may not survive the first problem.

The question concerning the number of petals in the graph of $r(t) = \sin(nt)$ is a nice example of an "extendable" problem. Using a graphics facility, students will quickly observe that there are n petals when n is odd and $2n$ petals when n is even, and this will probably be their report at the second meeting. More effort and coaching will be involved in attempting to explain this result analytically and will eventually lead to the recognition that there are always $2n$ petals, but each pedal is traced out twice when n is odd. Obtaining this result could be the follow-up problem or it could be worked out during the second meeting depending on the intellectual strength of the students. A next follow-up question could be to investigate $r(t) = \cos(nt)$. This will probably be an easy activity. If the students do not recognize that the plot of $r(t) = \cos(nt)$ is just a rotation of the plot of $r(t) = \sin(nt)$, then they should be led into discovering and explaining this result during the next meeting. A related question now is to determine the sectors which contain the petals. Another follow-up question, investigate the plot of $r(t) = \tan(nt)$, is more difficult. (There are no petals.) In particular, determine the order in which the branches are drawn as t increases from 0 to 2π. At the next meeting, note that $\pi/2$ is a vertical asymptote in rectangular coordinates and wonder what happens in polar coordinates. Are there asymptotes in polar coordinates? What would be a nice follow-up question at this point?

Faculty members who coach students in this type of activity should publicize their student's work to the rest of the department and to the admissions people who can use the information in their recruiting. Hearing a faculty member boast of the mathematics some students are doing on their own would certainly be a welcome relief from the student bashing that characterizes too many faculty conversations. The students should be provided with opportunities to share or show off their work. A special departmental colloquium, MAA Student Chapter program, undergraduate colloquium program at neighboring colleges, or a session of short Math Talks during Parents' Weekend are some of the opportunities that could be made available for student presentations. In addition to providing good speaking experiences for the students, these sessions are good public relations for the mathematics department, and, most important for the subject of this paper, they strengthen the case for using technology in doing mathematics.

Engaging students in doing mathematics outside the restrictions of a formal course has lots of advantages in addition to being fun to do. Most students view the individual attention as a compliment, and thus the activity enhances the confidence and self-respect of the students. The experience is a good forerunner for undergraduate research. In addition to the fun of involving students in doing mathematics, the activity provides the instructor with the opportunity to develop student assistants, i.e., lab assistants, paper graders, tutors, etc. Students who get turned on to doing mathematics without the carrot of a grade are very valuable to the health of a mathematics department. Their interests can be infectious with other students, and they are the students who give a mathematical meaning to MAA Student Chapters and math clubs.

2. Colleague Prong

Convincing colleagues to use technology in their teaching is the most important, and most difficult, aspect of the implementation stage. Most important because, without the commitment of the faculty to use technology and make the resulting appropriate pedagogical changes, the calculus program will probably not change. Alas, the department's statement of goals will be a fine sounding, but hollow, document. Difficulties abound, however. Computer anxiety, unwillingness to give up research time for calculus preparation, unwillingness to relinquish the security and ego aspect of lecturing, politics of gaining tenure, negative connotations associated with being involved with mathematics education, and the academic reward system are some of the difficulties encountered. However, the game is far from lost. The previous steps of building awareness and developing a syllabus for a reformed calculus program provide a strong base from which to move forward in the efforts to reform the calculus program—a program that emphasizes the use of technology as an exploratory tool as well as a computational tool, emphasizes active student engagement, and focuses the course on student learning rather than instructor lecturing.

An immediate objective of a program to encourage colleagues to use technology is to develop an advocacy group in order to avoid having the effort be viewed as a one or two person program. Although it is crucial to eventually involve well-respected, tenured members of the department in the group, it is expedient to start with the persons who are the most receptive. The following outlines a one-on-one strategy for developing an advocacy group for using technology in teaching mathematics:

- Breaking the Ice

  Sharing a neat discovery is an effective way to break the ice with the person you are trying to recruit. I remember the enthusiasm of the person who showed me that under magnification a small piece of the graph of a differentiable function appears to be a straight line giving credence to the term *locally linear.* I also remember my excitement when I first realized that I could approximate the solutions to an equation by first transforming the problem into one of finding the zeros of a function and then doing that by graphing. The enthusiasm of a person who has just made a discovery can be very contagious. In a similar manner, sharing the joy and satisfaction of guiding a student to a discovery can be very inspiring to colleagues.

- Eliciting Ideas

  People in general are complimented when asked for an opinion or an idea. Thus, asking the person you wish to recruit to sit in and critique a class in which you are using technology is a nice way to offer a compliment, gain valuable feedback, and involve your colleague. Additional discussions asking for feedback, particularly on pedagogical issues of technology-based activities, further deepens the involvement of your colleague. The emphasis of these discussions should gradually change from asking for feedback to eliciting ideas from your colleague. This could start with seeking ideas for exercises and out of class student problems and then expand into curriculum project development. The central objective is to move your colleague from a responding (offering feedback) stage into an active initiating stage and to do it in the positive context of improving instruction.

- Training

  If your colleague needs help in developing CAS expertise, then asking for a critique of a CAS "getting started" activity such as those found in [12] is a non-threatening transition into a training stage. Remember that being understanding of computer anxiety and alert to common beginning frustrations (i.e., How to log on? Off?) is very important in building self-confidence and to your overall goal of recruiting. Share your resources (problem sets, labs, papers, reports, etc.), but be careful not to overwhelm. In particular make sure your colleague has copies of Priming the Calculus Pump: Innovations and Resources (CRAFTY report), [16], and The Laboratory Approach to Teaching Calculus, [5]. Discussing an article or jointly working through a lab or "What iffing" a problem is a nice way of assisting a person to get into a resource. Inviting your colleague to join you in attending a regional or national CAS Workshop or a calculus minicourse at a national mathematics meeting may strengthen the person's sense of getting on a national bandwagon. A reasonable short term objective is to assist your colleague in developing and presenting a technology-based class topic.

- Advocating

  When the core group has grown to four or more, including a well-respected, tenured faculty member, it is time to organize for advocacy. Part of the program should be to establish a clear, public presence of using technology in teaching. Coffee pot discussions, lunch meetings, sessions for making up exercises, sessions for critiquing national recommendations, sessions for critiquing pedagogy, etc. are some useful activities that could be organized. Maintaining communication with colleagues in other disciplines, administrators, computer services, admissions are very important. In particular, I strongly recommend involving the dean in some of the sessions. An important, and useful, tactic in expanding the size of the advocacy group and in obtaining widespread support is to insist that all group sessions are open and to actively encourage outside people to attend.

3. Course Prong

There are several correct ways to introduce the use of technology into a calculus course. The spectrum of successful beginnings stretches from having one or two activities to almost daily use of a CAS or graphing calculator in the classroom. Many instructors have replaced a lecture class with a laboratory, while others have a calculator or computer linked to a projection device in the classroom, [3, 16]. There is a growing body of technology-based experience that is being shared through special sessions at professional meetings, and there are several resource materials available (problem sets, lab manuals, tutorials, etc), [4, 6, 12, 13, 15]. A word of caution however: it is dangerous to expect canned labs that are effective in one school to be equally effective at another school with a different instructor and different local idiosyncrasies. I encourage instructors to individualize exercises, labs, and projects found in the resource materials to suit their own courses and their own schools.

References [5, 16] should be required reading for anyone using technology in their teaching. A few introductory comments that have been cited by several persons are:

- The first activity in which students use a CAS or graphing calculator should be designed to develop student confidence in the use of technology.
- Two students working together is optimal.
- The instructor should be prepared to experience his or her role changing from being a presenter of facts to being a guide or a coach.
- Student lab assistants should be selected on the basis of their ability to communicate as well as their expertise in using a CAS.
- Most students need encouragement and guidance into exploring open ended questions. (They have little or no experience to guide them.)
- The instructor should be quick to divert questions to classmates and slow to give answers.

4. Equipment Prong

Equipping and maintaining computer labs is expensive. The institutional support gained through the activities carried out in the awareness step and through continuing open communications with administrators and computer services provides for a cooperative effort to obtain the necessary funding. Several foundations make computer equipment grants. A few examples are: the National Science Foundation through its Instrumentation and Laboratory Initiative program, the Fund for Improvement of Post-Secondary Education (United States Department of Education), the Carnegie Corporation of New York, the Exxon Foundation, and the Ford Foundation. Besides receiving financial assistance, receiving an outside grant adds legitimacy to the whole program of using technology in teaching mathematics. In addition to purchasing hardware and software, administrators need to understand the need for trained support personnel to maintain and service equipment. (Trying to use malfunctioning equipment is the surest way to frustrate and turn off students and faculty.) Faculty members should not be expected, or allowed, to take on this support function.

These four prongs are both interrelated and interdependent, like a double team of horses pulling a sled. The positive movement of one horse (prong) will encourage similar movement in the others, while the recalcitrance of one can be overcome by movement of the others.

Summary

Changing attitudes and creating acceptance among colleagues for activities that require more effort with only intangible rewards are difficult tasks. Success comes through patience, persistence, understanding, and careful planning. Creating an awareness among a broad segment of the institution of the need for change is crucial to gaining widespread support for reforming the calculus program. Within the department it is necessary to develop consensus on a syllabus that provides time for innovations and individualized topic treatment. Viewing the process of developing consensus as team building will lead to a supportive environment that is conducive to change.

Clearly stated course goals are absolutely necessary. They provide direction and focus to the implementation of the syllabus. The use of technology and the resulting changes in pedagogy will revitalize teaching and, in turn, enhance learning. Finally success ultimately depends upon effective communication with colleagues and administrators and the development of an advocacy group that includes well-respected, tenured members of the department.

References

1. *A Challenge of Numbers,* National Academy Press, Washington, D.C., 1991.

2. Douglas, R.G., editor, *Toward a Lean and Lively Calculus,* MAA Notes, Volume 6, Mathematical Association of America, Washington, D.C., 1986.

3. *Everybody Counts: A Report to the Nation on the Future of Mathematics Education,* National Academy Press, Washington, D.C., 1989.

4. Fraga, R.J., editor, *Calculus Problems For a New Century,* MAA Notes, Volume 28, Mathematical Association of America, Washington, D.C., 1993.

5. Leinbach, L.C., editor, *The Laboratory Approach to Teaching Calculus,* MAA Notes, Volume 20, Mathematical Association of America, 1991.

6. Jackson, M.B. and Ramsay, J.R., editors, *Problems For Student Investigation,* MAA Notes, Volume 29, Mathematical Association of America, Washington, D.C., 1993.

7. *Mathematical Sciences, Technology, and Economic Competitiveness,* National Academy Press, Washington, D.C., 1991.

8. *Moving Beyond Myths: Revitalizing Undergraduate Mathematics,* National Academy Press, Washington, D.C., 1991.

9. *Renewing U.S. Mathematics: Critical Resource for the Future,* National Academy Press, Washington, D.C., 1984.

10. *Renewing U.S. Mathematics: A plan for the 1990s,* National Academy Press, Washington, D.C., 1990.

11. *Reshaping School Mathematics, A Philosophy and Framework for Curriculum,* National Academy Press, Washington, D.C., 1990

12. Small, D.B. and Hosack, J.M., *Explorations in Calculus with a Computer Algebra System,* McGraw-Hill, Inc., 1991.

13. Solow, A.E., editor, *Learning by Discovery,* MAA Notes, Volume 27, Mathematical Association of America, Washington, D.C., 1993.

14. Steen, L.A., editor, *Calculus for a New Century,* MAA Notes, Volume 8, Mathematical Association of America, Washington, D.C., 1987

15. Straffin, P. D., editor, *Applications of Calculus,* MAA Notes, Volume 30, Mathematical Association of America, Washington, D.C., 1993.

16. Tucker, T. W., editor, *Priming the Calculus Pump; Innovations and Resources,* MAA Notes, Volume 17, Mathematical Association of America, 1990.

Graphing Calculator Intensive Calculus: A First Step in Calculus Reform for All Students

Bert K. Waits and Franklin Demana
The Ohio State University

Introduction

Computer generated numerical, visual, and symbolic mathematics is revolutionizing the teaching and learning of calculus. The computer can be a desktop computer with computer algebra and graphing software or a pocket computer with built-in software (graphing calculator). The content of calculus is changing—less time is spent on paper and pencil methods and more time is spent on applications, problem solving, and concept development [7, 9, 10]. And teaching methods are also dramatically changing—moving toward an investigative, exploratory approach.

We believe that graphing calculators are the appropriate computer tools for most students today (1994) because they are inexpensive (some less than $50), user-friendly, powerful (some built-in software on newer graphing calculators like the TI-85 and HP-48 is phenomenal), small, and personal. In short, a graphing calculator intensive approach is implementable for all students. Expensive and logistically complex computer laboratories are not necessary to teach a computer intensive calculus course. Any classroom today can become a computer laboratory with student use of graphing calculators. [1]

The Calculator and Computer Enhanced Calculus (C^3E) calculus reform project

We approach the incorporation of hand-held computer technology in calculus as a natural evolution of our positive experience in the large scale implementation of hand-held technology with all students in two previous projects: first, with calculators at Ohio State in the seventies [11] and then with graphing calculators in our highly regarded C^2PC project in the eighties [3]. Our C^2PC textbook, *Precalculus Mathematics, A Graphing Approach* is recognized as being the first widely adopted high school and college textbook to require graphing technology and is now in the third edition [6].

Our C^3E calculus reform project is based on what we learned in our many years with the C^2PC project. Fundamentally we learned that the principle of incremental change should guide our approach to calculus curriculum reform and the related integration of computer technology. We take a familiar body of calculus material and make the assumption that *every* student has an inexpensive, user-friendly *graphing calculator* for both in-class activities and for homework. We use graphing calculators as scientific calculators (they are the best we have ever used), as "tools" for computing derivatives and integrals numerically, as computers for programming (certain "tool box" programs like Simpson's method and Euler's method), as numerical "solvers" (e.g., root and intersection finders), and for computer visualization using their built-in graphing software (e.g., graphing derivatives, functions defined by integrals, and power series).

We believe technology will not be routinely used by all calculus students (or required by professors) until it costs less than $100, is user-friendly, and fits in a backpack or purse. The C^3E materials are reflected in a new textbook, *Calculus: Graphical, Numerical, Algebraic* [8] which requires graphing calculators. Our project does not assume that every student has a computer algebra system (like *DERIVE*™ or *Mathematica*®). However, in a few years, computer algebra will no doubt become much cheaper, and their use may then be a reasonable assumption. Colleagues who become comfortable with graphing calculators today will easily make the transition to the powerful and no doubt inexpensive computer algebra systems of the future [2].

Our philosophy of using graphing calculator numerical and visual methods to enhance the teaching and learning of calculus can be summarized by the following three points.

I. Do analytically (paper and pencil), then SUPPORT numerically and graphically (with a graphing calculator).

II. Do numerically and graphically (with a graphing calculator), then CONFIRM analytically (with paper and pencil).

III. Do numerically and graphically, because other methods are IMPRACTICAL or IMPOSSIBLE!

We are also convinced that required student use of graphing calculators today promotes a cooperative learning environment where calculus can be presented as an exciting, lively subject where student investigations become routine.

We illustrate our three point philosophy with four examples.

Point I: Use graphing calculators to visually support results first obtained by analytic calculus paper and pencil manipulations.

These "support graphically" activities are part of the "bread and butter" of a first step towards calculus reform. Here we take old familiar topics and support them with technology and, at the same time, we add to students' intuitive understanding of calculus concepts.

Problem 1 Use the limit definition to show that $\lim_{x\to 1} f(x) = -1$ where $f(x) = x^2 - 2x$.

The formal limit definition has remained a very mysterious concept to students. In fact, it is not commonly taught to first year calculus students in many universities. The limit concept can be dramatically enhanced by computer graphing and related numerical analysis. Analytically it can be determined that given any $\epsilon > 0$, choosing $\delta = \sqrt{\epsilon}$ will satisfy the usual limit definition when applied to Problem 1. Here a graph can be much more instructive. The graph of $f(x) = x^2 - 2x$ (Figure 1) clearly indicates the continuous nature of the function (so the limit at $x = 1$ can be calculated by evaluating $f(1)$). However, the analytic "limit proof" of this fact is not so clear.

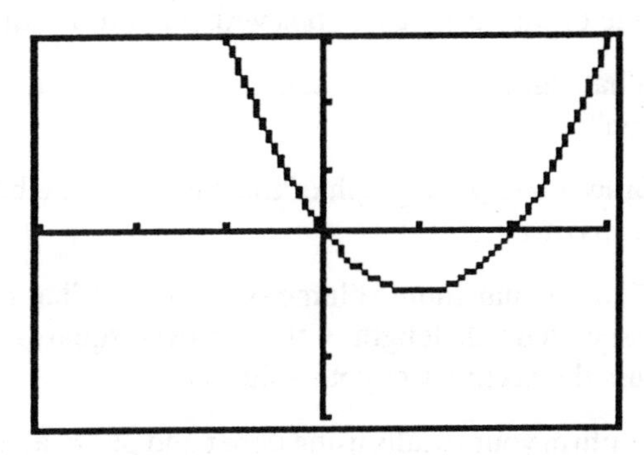

Figure 1: The graph of $f(x) = x^2 - 2x$.

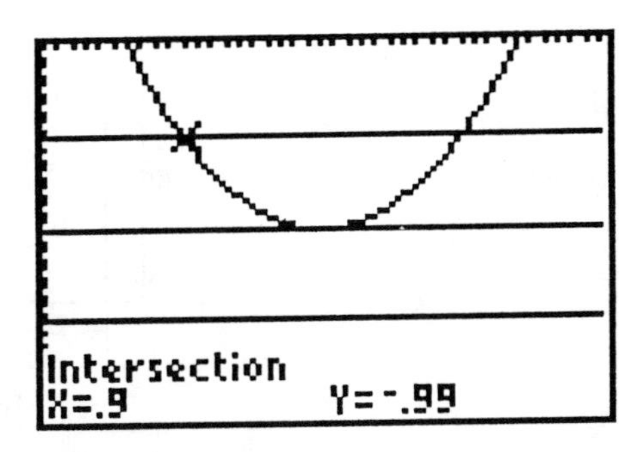

Figure 2: The graph of $f(x) = x^2 - 2x$, and the lines $y = -1, y = -1 - .01, y = -1 + .01$ for $0.8 \leq x \leq 1.2$ and $-1.02 \leq y \leq -0.98$

A computer generated "magnified" graph is very valuable. In Figure 2, we illustrate the limit definition for a "given ϵ" of $\epsilon = 0.01$. The student adds the "target" lines $y = -1 \pm \epsilon$ or, in this case, $y = -1 - .01 = -1.01$ and $y = -1 + .01 = -0.99$. Then a graphing ZOOM procedure is used to obtain the graph shown. It becomes clear that if x is kept between 0.9 and 1.1, the function values $f(x)$ are always between -1.01 and -0.99. That is, if $|x - 1| < \delta = 0.1$, then $|f(x) - (-1)| < \epsilon = 0.01$. This fact strongly suggests that $\delta = 0.1 = \sqrt{0.01} = \sqrt{\epsilon}$ is the required delta value in terms of epsilon in the analytic limit analysis.

This example is very easy to deal with because the limit point was at the minimum value of the function. The *principle of local linearity* will help for other values. For example, suppose the problem is changed to "Use the limit definition to show that $\lim_{x\to 1.5} f(x) = -0.75$." Figure 3 shows a magnified computer ZOOM-IN view of the graph in Figure 1 at the point $(1.5, f(1.5))$ and at an-

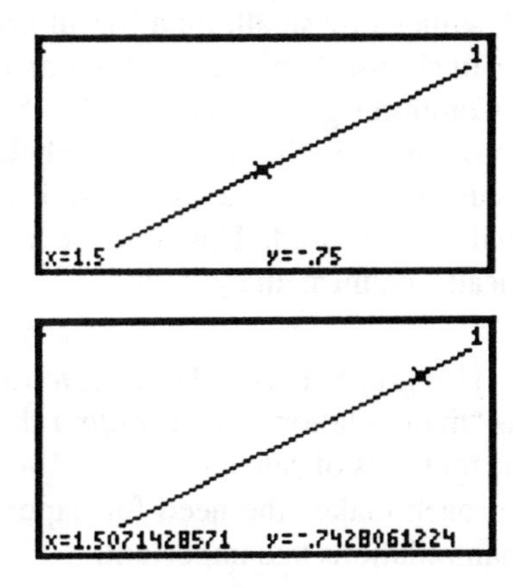

Figure 3: Zoom-in views of the graph of $y = x^2 - 2x$ near the point $(1.5, -0.75)$.

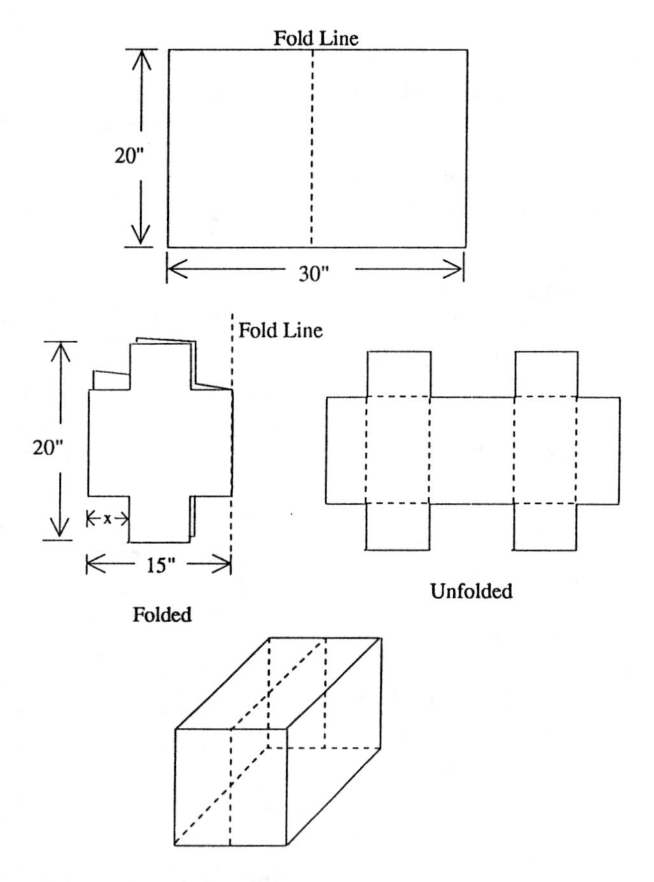

Figure 4: Constructing the "brief case" box with lid.

other nearby point. Notice that the graph of $f(x) = x^2 - 2x$, for all practical purposes, is a straight line with slope $m = (-0.75 - (-0.7428061224))/(1.5 - 1.5071428571)$ which is very close to 1 (actually 1.00714285715). That is, the function $y + 0.75 = x - 1.5$ or $y = x - 2.25$ closely approximates the quadratic function near $x = 1.5$. The fact that the slope has value 1 strongly suggests that for any $\epsilon > 0$, sufficiently small, then for all practical purposes, we can choose $\delta = \epsilon$ in the limit analysis of this example. Computer graphing can make the limit definition far more meaningful than past analytic paper and pencil "hocus-pocus." One can go on and complete the analytic analysis if desired. This could be an example of "do graphically, confirm analytically."

Point II: Use graphing calculators as *tools* to actually *do* calculus "manipulations" then *confirm* the results using analytic methods of calculus.

This approach makes the need for paper and pencil calculus manipulations less important.

Problem 2 The "brief case" box with lid problem.

A box with lid is constructed from a 20 by 30 inch sheet of material in the following manner. First the material is folded in half forming a 20 by 15 inch double sheet. Then four equal squares of side-length x are removed from each corner of the folded sheet. The material is then unfolded and a box with sides and a lid are formed by folding along the dotted lines shown in Figure 4.

The following questions are typical of the investigations we ask students to deal with routinely.

a. Determine an algebraic representation of the volume of the "brief case" box with lid in terms of x.

b. What values of x make sense in this problem situation?

c. Draw a complete graph of the volume of the box in terms of x.

d. Find the maximum volume of the box. What is the associated side length of the removed square? Discuss the accuracy of your solution?

e. Confirm your results using paper and pencil analytic methods of calculus.

Figure 5A displays the graphs of the function $y = V(x) = 2x(15 - 2x)(20 - 2x)$ and its first derivative (using the numerical derivative feature). The graph of $y = V(x)$ for $0 < x < 7.5$ is the graph of the box volume problem situation. The figure also shows that a "solver" (for example, ROOT on the TI-82 or TI-85) has been applied to the derivative graph to find the zero of the derivative. Figure 5B shows the function value at the "root" of the derivative (which is the local maximum value of $y = V(x)$). Thus the student can apply the theory of calculus ("look for possible local extrema where the derivative is zero") and solve this problem to a very high degree of accuracy using a graphing calculator. Students can then be required to CONFIRM the result analytically by computing the derivative and solving the resulting equation using ordinary paper and pencil calculus. A lively discussion will ensue when students are asked to write about and contrast both solution methods.

An interesting related exercise involving a "simple" equation is given by the following variation of problem 2.

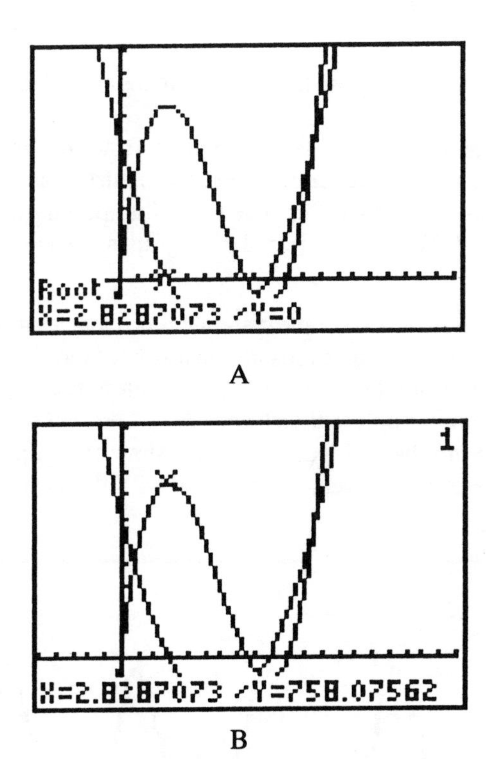

Figure 5: Part A. The graphs of $y = V(x)$ and $y = V'(x) = \text{NDer}(V, x)$. Part B. The (local) maximum value of $y = V(x)$ is $V = 758.07562$ (accurate to all digits shown).

Find the side length of a removed square to obtain a brief case box with volume 455 cubic inches. Can you confirm your results using analytic methods? If so, do it! Can you find the exact solution? If so, do it! Write a paragraph outlining the issues involved in applying an analytic method versus applying a numerical/graphical method.

Point III: Solve easily stated and understood problems that calculus students can't solve with paper and pencil analytic methods. And some that have *no* analytic solution.

Illustrate mathematical ideas and applications in concrete geometric settings. We explore, investigate, and make and test mathematical conjectures.

Problem 3 Visually illustrate the Fundamental Theorem of Calculus. [5]

Background Consider a continuous function f defined on an interval $[a, b]$ (*any* continuous function even those without closed form antiderivatives). The Fundamental Theorem of Calculus guarantees the existence of a function F, namely $F(x) = \int_a^x f(t)\,dt$, with the property that $F'(x) = f(x)$. The problem is that, until today, students could "find" these antiderivatives that we know exist for only a relative few *contrived* functions f. And these contrived functions are what make up typical calculus textbook integration problems! However, today with graphing calculators all students can "see" the antiderivative F easily for any continuous function (and those with continuous extensions) even if we can't write the explicit "closed form" analytic expression. All that is needed is a way of graphing $F(x) = \int_a^x f(t)\,dt$. The TI-82 and TI-85 have this as a built-in feature. Other graphing calculators can be programmed with this feature as a "tool box" item. See the *Technology Resource Manuals for Calculus* that accompanies our textbook for programs for the TI-81, Sharp 9200 and 9300, Casio, and Hewlett Packard graphing calculators [4].

Solution The Fundamental Theorem of Calculus implies that

$$D_x[F(x)] = D_x\left[\int_a^x f(t)\,dt\right] = f(x).$$

This is a TI-82 or TI-85 activity. Graph the function $y = \text{FnInt}(t^2, t, \{-2, 0, 2, 3\}, x)$ in the $[-5, 5]$ by $[-10, 10]$ window ($-5 \le x \le 5$, $-10 \le y \le 10$). This produces four graphs of $F(x) = \int_a^x t^2\,dt$ for $a = -2, 0, 2, 3$. Note the use of a *list* in the lower limit of integration position. Students can conjecture about what are the analytic forms of the antiderivatives that they "see." And they can test their conjecture by "overlaying" the analytic expression (e.g., DRAWF $\frac{x^3}{3} + C$). EXPLORE: Determine C for the above four antiderivatives and explain how C and a are related, etc.

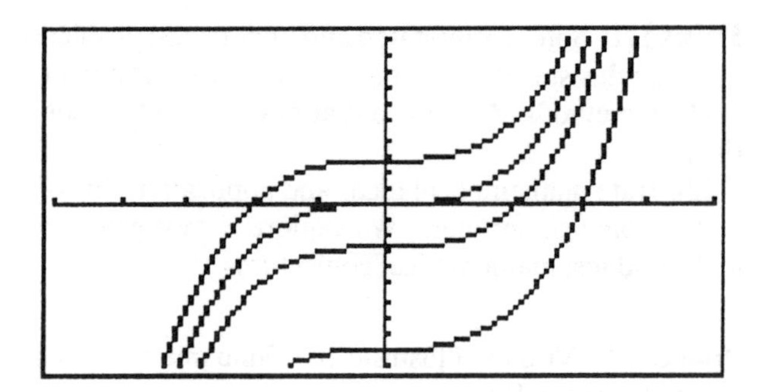

Figure 6: A graph of a family of antiderivates of x^2 in $[-5, 5]$ by $[-10, 10]$.

Figure 6 makes the constant of integration and family of antiderivatives concept come alive for students. Here are "connected" slope fields! Static figures do not do this activity justice. This dynamic activity must be "experienced" by the student.

Next we graph the numerical derivative of this function defined by the integral to visually illustrate the fundamental relationship that the derivative of $F(x)$ is $f(x)$ as claimed by the Fundamental Theorem of Calculus. That is, we show that $D_x\left[\int_a^x f(t)\,dt\right] = f(x)$ for *any* continuous function f. The graph of

$$y = \text{NDer}\left(\text{FnInt}\left(\frac{\sin t}{t}, t, 0, x\right), x\right)$$

on the TI-82 or TI-85 is the graph of

$$y = D_x\left[\int_a^x \frac{\sin t}{t}\,dt\right]$$

which should be $y = \frac{\sin x}{x}$. Figure 7 shows this is indeed the case! Note: we regard the integrand to be the continuous extension of $\frac{\sin x}{x}$. TRACE can be used to compare the two function values for supporting numerical evidence. This activity is a powerful visualization!

The predator-prey problem The classic Volterra predator-prey problem becomes a routine exercise using a graphing calculator like the TI-85. The model assumes the rates of population growth of predator-prey populations (foxes and rabbits) are related by the "highly coupled" first order differential equation system given by

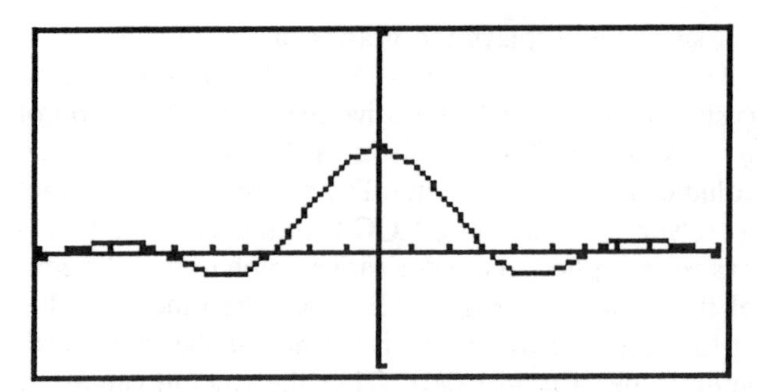

Figure 7: Graphs of $y = d_x\left[\int_a^x \frac{\sin t}{t}\,dt\right]$ and $y = \frac{\sin x}{x}$ in $[-10, 10]$ by $[-1, 2]$ appear to be the same.

$$\begin{aligned} dF/dt &= (-0.5 + 0.02R)F \\ dR/dt &= (1 - 0.1F)R \end{aligned}$$

where $y = F(t)$ is the population of foxes at time t and $y = R(t)$ is the population of rabbits at time t (t measured in years).

Problem 4 Suppose there are 10 foxes and 50 rabbits at time $t = 0$ (today).

What are the population graphs? How are the populations related over time? (Here a picture is as good as an analytic result!) Suppose the initial populations are changed. How do the population graphs change?

Solution This simple system has no closed form solution. Numerical methods are necessary. Here is a TI-85 solution using its amazing built-in differential equation solver with graphics interface.

The graphs in Figures 8 and 9 show the population graphs over a 40 year time period with the given initial

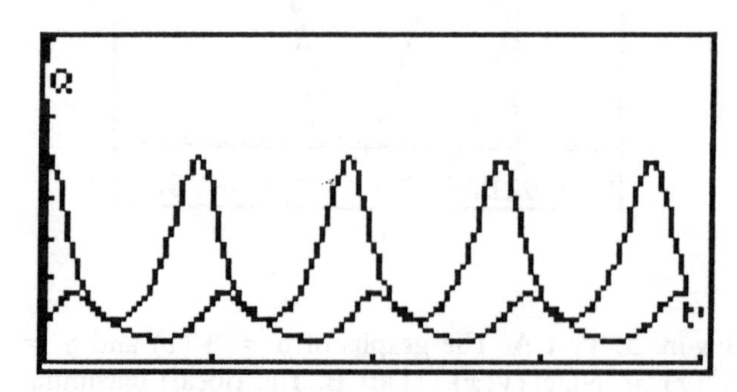

Figure 8: The fox and rabbit populations for 40 years starting with 10 foxes and 50 rabbits.

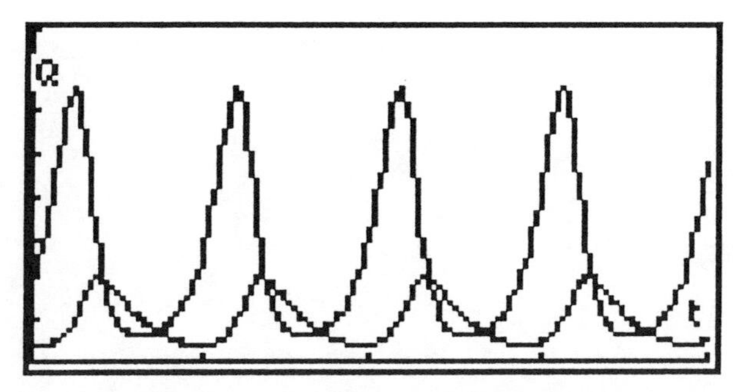

Figure 9: The fox and rabbit populations for 40 years starting with 4 foxes and 20 rabbits.

populations and with a change in the initial populations to 4 foxes and 20 rabbits.

Figure 10 shows the population patterns for various initial conditions in one view (orbits—the phase plane solution). These orbits are found by plotting the points $(F(t), R(t))$ for 6 different initial conditions (the beginning populations) which the TI-85 does automatically with an axes change selection.

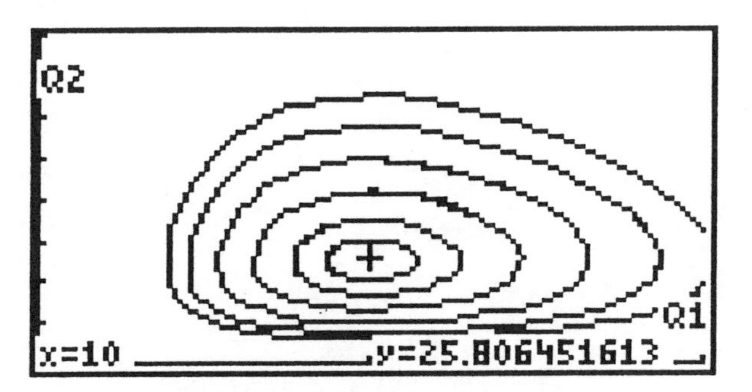

Figure 10: A phase plane solution to the fox-rabbit predator-prey problem.

The student can be led to conjecture that perhaps there is a set of initial conditions that result in stable (constant) populations over time. The graphs suggest that if the starting population is 10 foxes and 25 rabbits then the populations will be stable. This result can then be confirmed analytically using paper and pencil calculus (when does $dF/dt = 0$?, etc.) and supported graphically.

Summary

Hopefully we have made the case that the types of graphing calculator activities represented by the four problems in this paper provide insight and understanding for all students in ways not possible with paper and pencil methods alone. These types of activities will also empower and excite calculus students in ways not possible with paper and pencil methods alone.

We readily admit that we have not yet made all the hard decisions regarding what content should be modified or deleted from the current calculus curriculum. The jury is still out! However, we believe our C^3E project is an important first step. Indeed, it is a necessary first step, a step of enhancing the traditional calculus curriculum with computer numerical and visual methods delivered by student use of inexpensive graphing calculators.

We are also able to include many rich examples and illustrations that are possible for all students only with graphing calculators. Furthermore, graphing calculators promote sound mathematics teaching methods including cooperative learning, student investigations, and writing about mathematics.

References

1. Demana, Franklin, and Bert K. Waits. (1992) A computer for all students. *The Mathematics Teacher,* 84(2).

2. Demana, Franklin, and Bert K. Waits (1992) A case against computer symbolic manipulation in school mathematics today. *The Mathematics Teacher,* 84 (3).

3. Demana, Franklin, and Bert K. Waits. (1994) The calculator and computer precalculus projects (C^2PC): What have we learned in ten years? in *Impact of Calculators on Mathematics Instruction.* Monograph, University of Houston. George Bright (Ed.). Lanham, Maryland: University Press of America, Inc.

4. Demana, Franklin, and Bert K. Waits. (1994) *Technology Resource Manuals for Calculus.* Reading, MA: Addison-Wesley Publishing Company.

5. Demana, Franklin, and Bert K. Waits. (in press) *The Power of Visualization in Calculus.* TICAP Resource Guide. Monograph, Clemson University. Harvey, John and John Kenelly (Eds.).

6. Demana, Franklin, Bert K. Waits, and Stanley Clemens. (1994) *Precalculus Mathematics, A Graphing Approach, Third Edition.* Reading, MA: Addison-Wesley Publishing Company.

7. Douglas, R. (Ed.). (1986) *Toward a Lean and Lively Calculus: Report of the Conference/Workshop to develop curriculum and teaching methods for calculus at the college level.* MAA Notes no. 6. Washington, DC: Mathematical Association of America.

8. Finney, Ross, George Thomas, Franklin Demana, and Bert Waits. (1994) *Calculus: Graphical, Numerical, Algebraic.* Reading, MA: Addison-Wesley Publishing Company.

9. Steen, Lynn A. (ed.). Mathematical Association of America. (1987) *Calculus for a New Century.* MAA Notes no. 8 Washington, DC: Mathematical Association of America.

10. Tucker, Tom. (ed.) Mathematical Association of America. (1990) *Priming the Calculus Pump: Innovations and Resources.* Prepared by the CUPM subcommittee on Calculus Reform and the First Two Years. MAA Notes no. 17. Washington, DC: Mathematical Association of America.

11. Waits, B. K., and J. Leitzel. (1976) Hand-held calculators in the freshman mathematics classroom. *American Mathematical Monthly,* 83(9).

The Interface between Precalculus and Calculus

Calculus Reform: The Need for Dialogue with Reforms at the Secondary Level

Lyle Andersen and Maurice Burke
Montana State University

Introduction

The findings established by a national survey for secondary mathematics reform are discussed in this paper. These findings are consistent with the changes proposed by the NCTM *Standards.* This report also addresses the anticipated changes in the secondary mathematics curriculum, how these changes will interface with the calculus reform content and pedagogy, and what major obstacles preventing calculus reform appear to stand in the way at many of the universities throughout the country.

Background For Supporting the Secondary Mathematics Reform in Montana

In the fall of 1988 in order to ascertain the extent of interest in, and the implications of, adopting an integrated mathematics program and to develop a policy report regarding its implementation, a consortium of mathematics educators from Montana and Washington requested and received a grant from the Exxon Foundation to conduct a national survey on these issues. A questionnaire was developed in the Fall of 1988 and distributed nationally in March 1989. This survey was also designed to determine the content, pedagogical strategies, and expected outcomes of an integrated secondary mathematics program. The survey included all 50 state mathematics supervisors, and a national random sample of 500 mathematics supervisors, 500 mathematics teacher educators, and 500 mathematics teachers. Results from 27 (54%) state supervisors, 140 (31%) district supervisors, 164 (33%) mathematics teacher educators, and 140 (28%) mathematics teachers were compiled in order to develop a first draft report. The results of the survey showed there is very strong support, from all groups, for movement to an integrated secondary mathematics program to meet the needs of all students [1, pp. 2–3].

This survey also led to a definition of an integrated mathematics curriculum. An integrated mathematics program for all students is a holistic mathematical curriculum which:

- consists of topics chosen from a wide variety of mathematical fields, and blends those topics to emphasize the connections and unity among those fields;
- emphasizes the relationships among topics within mathematics as well as between mathematics and other disciplines;
- each year, includes those topics at levels appropriate to students' abilities;
- is problem centered and application based;
- emphasizes problem-solving and mathematical reasoning;
- provides multiple contexts for students to learn mathematical concepts;
- provides continual reinforcement of concepts through successively expanding treatments of those concepts;
- makes use of appropriate technology.

Expected Outcomes

The survey also provided a clear mandate regarding expected outcomes [1, pp. 4–5].

Outcomes for students:

- Students from all ability levels will take more mathematics, with the greatest interest coming from those of average ability.
- Students will find mathematics more interesting, will have a greater understanding of mathematics, and will have less loss of skills over time.
- Students will be better prepared in mathematics with the greatest expectation for improvement in the noncollege-bound group.

- Student achievement scores on current standardized tests will not be adversely affected.

Outcomes for teachers:

- There will be greater communication among teachers using an integrated mathematics program than among those using a traditional program.

- Teachers will think of themselves as mathematics teachers rather than as algebra or geometry teachers.

- Teachers will teach topics from a broader range of mathematical fields using multiple contexts.

The SIMMS Project

As a result of this survey, a consortium consisting of the Montana Council of Teachers of Mathematics, the Montana State Office of Public Instruction, Montana State University, and The University of Montana was created which submitted Montana's State Systemic Initiative (SSI) Proposal to the National Science Foundation. The proposal was subsequently funded and is known as the Systemic Initiative for Montana Mathematics and Science (SIMMS) Project. The main objectives of SIMMS are the following:

- Redesign the 9-12 mathematics curriculum using an integrated, interdisciplinary (mathematics and science) approach for all students.

- Develop inservice and preservice programs and policies on integrated mathematics to prepare teachers of grades 9-16 and collaborate with similar efforts for grades K-8.

- Coordinate efforts among the science community to develop a state vision aligned with national standards in science reform.

- Develop the support structure for legislative action, public information, and the general education of the populace necessary to implement new programs.

- Increase the participation of females and Native Americans in mathematics and science.

- Act as agents of change in the area of assessment and provide on-going project assessment.

Comparing the Calculus Reform and the Secondary Mathematics Reform

In addition to the SIMMS project, the NSF has funded the development of four comprehensive high school integrated mathematics programs based upon the use of technology in real world applications. These are The Interactive Mathematics Project (San Francisco State University), Applications/Reform in Secondary Education (COMAP), Core-Plus Mathematics Project (Western Michigan University), and The Secondary Mathematics Core Curriculum Initiative (Hartford Alliance for Mathematics and Science Education). The high school mathematics curriculum changes being promoted by these and other projects will strongly support a wider acceptance of the calculus reform movement. Why? Simply stated, it seems obvious to us that students from these projects are going to opt for higher education programs using familiar pedagogical methods and materials. Furthermore, the calculus reform curricula being developed nationwide utilize the same major paradigm shifts as the secondary (grades 9–12) school mathematics curricula currently being developed: the traditional secondary and college mathematics curricula are mostly about *answers*, while reform curricula at both levels (secondary and calculus) are about *questions*.

These secondary mathematics reform projects and the calculus reform projects stress (1) *applications* of the content to "real world" situations, (2) hands-on *active learning* strategies, rather than the traditional lecture method, (3) *group learning* in both in-class activities, and out-of-class projects, (4) students *writing* about mathematics in order to deepen conceptual understanding, (5) *technology* such as graphing calculators and computer software as an integral part of the teaching of mathematical concepts, and (6) significant changes in the way assessment is done. All de-emphasize symbolic manipulation. Graphing calculators and computers do this quite well.

The similarities in pedagogical and content changes in the secondary preparation curricula and the reform calculus curricula lead us to believe that there are grounds for productive dialogue between the respective levels. There are important issues that such a dialogue could clarify. For example, many educators accept the analogy that symbolic manipulators (e.g., *Maple, Mathematica*®, *DERIVE*™) are to the calculus as calculators (e.g., TI-82) are to arithmetic. What policies should be adopted concerning the use of symbolic manipulators in the reform calculus and in secondary level mathematics? Should the NCTM and/or MAA take a position on this issue? These are important questions to the SIMMS project, which is building the use of symbolic manipulators into its high school curriculum.

Obstacles to Reform

In order for the secondary mathematics reforms and the reforms of calculus and other related courses to be successful, it is necessary to develop administration and teacher awareness of, and support for, the changes required.

First, the traditional university attitude towards mathematics of "stack 'em deep and teach 'em cheap" must be changed. This prerequisite to reform at universities such as ours cannot be overemphasized. Large class size, little or no technology or financial support and bolted down desks instead of tables are some of the barriers we face.

Unless we are able to convince our administration that we can't afford not to change because of what is happening at the secondary level, there is little chance of calculus reform courses being anything more than an anomaly. Presently, mathematics classes at our university, including calculus, average nearly 40 students per class at a dollar cost of about $1600 per student FTE. The University's lab science departments (Biology, Physics, and Chemistry) have a cost per student of from $2100 to $3300 per student FTE. The lab science courses have a ratio of student FTE to faculty FTE of approximately half of what we have in mathematics. For us the direction we should move is obvious, but how to convince the administration that mathematics should become more of a laboratory science seems incredibly difficult. The problem may be simplified by the fact that most of our first year calculus is taught by graduate teaching assistants who, we believe, are more apt to go for a reform calculus program.

Second, changing mathematics courses from lecture-type classes to laboratory-based classes highlights what may be the most obstinate barrier. That barrier is faculty attitude. Overcoming this barrier requires a significant change in whom the faculty perceive are the clients of college calculus. Nationwide, many see the calculus as a filtering service for other departments which are considered to be the clients. Others are looking for "a few good men" being trained for the advanced mathematics classes, and it is these students who are perceived as the clients. It is easy to be content with the lecture-type calculus class if one holds these attitudes. When faculty start to view *all* students as clients, then the high failure and drop-out rates of the lecture-type class will lead to them considering reform calculus options.

A third obstacle to reform is the lack of dialogue between those interested in reforming the secondary mathematics curriculum and those involved with early undergraduate level mathematics, including the calculus. This lack of dialogue reduces the sense of urgency in university administrations and mathematics faculties to make any changes in the calculus. In turn, this failure to consider reforming the calculus often results, for example, in college placement exams designed for traditional calculus courses, which become barriers to students coming from the new secondary school curricula.

Conclusion

Two important issues must be addressed if informed changes are to occur. They are (1) college administration support along with faculty involvement and support, and (2) interfacing the secondary curriculum with the university mathematics curriculum.

Administrative support regarding reform is critical. It should be noted, for example, when universities, such as the University of Michigan, have received administrative support for calculus reform significant progress has resulted. We do not anticipate being able to initiate significant changes of faculty responsibilities without the administration becoming vitally involved. It is not just a financial issue. Changing the traditional reward system for research faculty may, in the long term, be of even greater importance. Unless the reward system is changed, our mathematics faculty's attitude toward reform will likely continue as it has in the past. A few committed individuals will continue to step forward to make some changes in their courses, but the teaching role of the faculty will remain as it is now.

The lack of a purposeful plan for interfacing the secondary mathematics curriculum with the university mathematics curriculum poses serious risk for our future students. If the college curriculum does not reflect the recommendations of the NCTM *Standards* which are presently being implemented into the secondary curriculum, then we will widen the gap between secondary mathematics and the college curriculum. This could lead to increased remedial mathematics programs. If, on the other hand, some of the secondary schools do not change their mathematics curriculum, then their students will be at a disadvantage if they attend colleges that have implemented significant course reforms in mathematics.

Bibliography

1. Dolan, Dan, Jack Beal, Johnny W. Lott, John P. Smith, *Integrated Mathematics; Definitions, Issues, Implications* (Executive Summary); Montana College of Teachers of Mathematics, University of Montana, Missoula (1989).

Probability: Connecting Concepts in Precalculus and Calculus

Martin E. Flashman

Mathematics Reform in the 1960s Compared to Today's Reform

The current new wave in mathematics is bringing changes in pedagogy, technology, and content that are reminiscent of reform during the late 1960's. At that time two major reforms were proposed, using computers and focusing on probability. Both failed to gain lasting widespread support.

Computers were suggested in the '60s as an aid for instruction, but the computer technology of the time focused on the computational power of computers. Programming was an essential element of this reform that aimed to give students control over the technology. One reason this reform failed was the time taken from calculus instruction to learn computer programming. Without students and faculty already programming literate, the time required to begin to use the computer was a serious disadvantage. A remnant of the technology wave continued into the '70s when programmable hand-held calculators were available. This suffered from similar time problems.

Today's technology revolution in the calculus and precalculus reform movements places much greater emphasis on the graphing capabilities of technology. Even though some programming is sometimes required, the level of detail and the necessity of mastering an unfriendly programming language have been reduced greatly by menus and more interactive features. It seems almost certain that with today's technology and the promise of even better technology in the future, this aspect of reform will become a part of the establishment.

The second major thrust of the reform movement of the 1960s was the connection of calculus with applications in probability and statistics. The rationale given for this reform was that the majority of students taking calculus (and mathematics in general) will encounter concepts from probability and statistics more frequently in their life activities than most topics found in the standard calculus course. The call for relevance of the '60s still resounds in the current wave of reform, and the rationale is perhaps stronger today than it was then.

Like the computer reform, the attempt to introduce probability and statistics in calculus did not succeed in the '70s for reasons of time economics. It was generally believed then that to teach probability concepts required a discussion of discrete combinatorial probability. An introduction to continuous (geometric) probability had to be postponed till after treating much non-calculus material.

Taking time from calculus for combinatorial concepts prevented the meaningful incorporation of probability into the calculus. Either students had be conversant with combinatorial probability before entering the calculus sequence or the calculus course would treat the probability as a more perfunctory application for calculus without the need to make it sensible in the calculus context. "They'll learn this later in statistics courses" was and still is a familiar rationalization for the "they can wait" position.

The current precalculus and calculus reform interest in probability seems to be following the same route of the '60s and '70s. The two California Frameworks and the NCTM *Standards* certainly take the position that probability and statistics are important strands in the fabric of the mathematical sciences. They call for these topics to be woven into the fabric of experience and understanding of all current mathematics students. Unfortunately the assumptions of the earlier reform still seem to dominate the approach taken to probability and statistics. For example, students in primary and elementary grade levels are supposed to have experience with both combinatorial and continuous random variables (such as foot size and heights). However, the high school curriculum focuses almost exclusive attention on discrete combinatorial situations, such as playing with dice and taking balls from urns.

Some calculus reform projects have included an occasional problem or a special section on probability. Though fine in their limited approach, these do not give probability the same status as the tangent, velocity, or area problems. They miss the opportunities that probability provides for illustrating many essential interpretation and application elements of the calculus.

The object of this paper is to suggest that probability can be treated in a more fundamental way in the precalculus and calculus curricula. Probability concepts can and should be connected to the concepts and techniques of these mathematics courses. Included below are some details for implementing this suggestion which the author has introduced in the Sensible Calculus Program at Humboldt State University. Actual materials from the Sensible Calculus program will be available for limited class testing sometime in 1994–96.

The Sensible Calculus Program Approach to Probability. Darts, Not Dice

The treatment of probability in the Sensible Calculus Program begins with a discussion of a model for a random event that will continue to be explored and generalized throughout the precalculus and calculus courses. It provides a blend of finite and continuous probability concepts and builds progressively towards calculus concepts. The precalculus introduction emphasizes the relation between probability distribution and density functions visually, numerically, and algebraically. This follows the sensible calculus program's general philosophy that precalculus instruction should give students motivation for further study in calculus through problem solving situations.

So what is the modeling situation? We consider the simple experiment of throwing a dart at a circular region assumed to have a unit radius. Although other choices for the basic experiment can work as well, darts seem to be a sufficiently common experience for students that they comprehend the situation quite readily.

From the beginning the experiment is distinguished from the questions and events that relate to the experiment. An initial question for students is to estimate what proportion of a large number of throws (say 100) will land within the concentric circle with radius $1/2$. After some discussion of the situation, students can conclude that the proportion here is related to the area of the regions and that the proportion is approximately $1/4$. This fact can be observed numerically by having a computer simulate the dart experiment 100 times. We also record the distance, R, that the dart falls from the center of the circle. In this way we encounter a random variable as a number that is measured in a repeatable experiment that helps resolve a question about the experiment. Through this example we have defined a function R from the experiments to the real numbers, and the range of the random real R is the interval $[0, 1]$. (Confusion over the fact that probabilities are numbers in this same interval is actually helpful to distinguish the numbers from their uses.)

After experience with the dart experiment and the random variable R, students can investigate more general questions such as determining the probability that the dart will fall within other regions of the circle. These lead to developing the notion of a probability distribution function. For this example the probability distribution function answers the following question: What is the probability that the dart lands within or on a concentric circle of radius A, where A is in the interval $[0, 1]$? Translating this question to random variable language, the function gives the probability that the random variable has a value less than or equal to A.

Many questions in probability can be explored once the distribution function is established. Questions about the probability of the random variable having certain properties replace questions about probabilities for certain situations happening in the experiment. Many of these questions can be explored and resolved partially in the precalculus course. For example, in our dart experiment, the distribution function is seen to be given by $F(A) = A^2$. This simple function allows one to discover and confirm using computer simulation that the median for this experiment is about $7/10$.

A related question is how to partition the circle using concentric circles so that the dart will fall in the resulting rings with about equal frequency for 100 throws. This can lead to a discussion at the precalculus level of estimating the average value of R that will result from 100 throws. This may even lead a class to investigate sums of squares and perhaps to discover the theoretical mean is $2/3$. Later work in calculus on the mean of a random variable using integration builds on this foundation.

Probability density is another concept that can be explored at a precalculus level. The dart experiment also works well for this. Focus on the probabilities of the dart landing in different concentric rings, or equivalently, the probabilities that the random variable R takes values in different intervals such as $[1/8, 1/4]$ and $[3/4, 7/8]$. Noticing different probabilities suggests comparing the probabilities to the interval lengths. This leads to the definition of the average probability density of a random variable over an interval as the ratio of the probability that R is in the interval to the length of the interval. For the dart experiment the average probability density of R for the interval [A,B] is easily seen to be

$$\frac{F(B) - F(A)}{B - A} = \frac{B^2 - A^2}{B - A} = B + A.$$

The complementary relation between probability distribution and density is easily recognized with this approach using the distribution function as the primitive concept. Later in the calculus course probability concepts of a random variable which have already been seen in the precalculus materials can be reviewed with the calculus concepts. Thus the median connects to continuity and the intermediate value theorem; the mode connects with extrema, critical points, and convexity; the relation between distribution and density relates differentiation, differential equations with boundary conditions, and the fundamental theorem of calculus; and the mean provides an excellent application of the relation of sums to integrals.

Other probability issues relate to a variety of calculus concepts including variation, exponential and normal distributions, estimation of integrals, polynomial estimation, and even volumes of revolution! Connections between the discrete and continuous are frequently possible, but do not require any extended combinatorial analysis.

Having a theme that can connect concepts and techniques of precalculus and calculus levels of learning can improve both courses. Though probability is not unique in supplying such a theme, it is a very relevant and important application of calculus. This paper has tried to suggest some ways in which the probability theme can be played to reinforce rather than distract from the central concepts of calculus. It complements other themes such as estimation, differential equations, and modeling. As I have suggested elsewhere in [1], [2], and [3], such themes can help make a more sensible calculus course. The conclusion should be clear: In future waves of precalculus and calculus reform the study of continuous probability should and will increase as a central theme for application and connecting concepts.

References

1. Flashman, Martin, "Editorial: A Sensible Calculus," *The UMAP Journal,* 11 (1990), pp. 93–95.

2. Flashman, Martin, "Differential Equations: A Motivating Theme for A Sensible Calculus," in The Report of A Conference on Calculus and Its Applications held at the University of Texas, San Antonio, NSF Calculus Reform Conference, October, 1990, to appear.

3. Flashman, Martin, "Concepts to Drive Technology," *Proceedings 5th Annual International Conference on Technology in Collegiate Mathematics,* November, 1992, Addison-Wesley, 1993.

Lessons from the Calculus Reform Effort for Precalculus Reform

Sheldon Gordon
Suffolk Community College

Deborah Hughes Hallett
Harvard University and University of Arizona

Mathematicians, scientists, and philosophers have long regarded calculus as one of the greatest intellectual achievements of western civilization. Unfortunately, very few of the students who have taken a traditional calculus course share this sentiment. All too often, they have instead seen calculus as:

- a collection of poorly understood rules and formulas involving manipulations,
- a set of artificial problems that provide little feel for the power of calculus,
- a collection of poorly understood and rarely appreciated theoretical results that were memorized and regurgitated.

It is not surprising that these courses have been unsuccessful for so many otherwise capable students.

Most of the calculus reform projects have attempted to address this problem by creating new and more motivating approaches to calculus. Most have focused on the applications of calculus. Many have addressed the problem through the use of sophisticated technologies. Some have considered innovative ways to deliver the mathematics, say in the context of group projects or in a laboratory environment. Others have considered the content of the course, especially when the students they address have had calculus in high school.

In this article, we will describe the philosophy behind the approach adopted by the Calculus Consortium based at Harvard (CCH) and outline the present status of the project. We believe that many of the themes we have adopted and our experiences may provide direction to those who wish to reform the courses that lead to calculus. Many of the same ideas certainly apply to the other calculus reform projects as well, and we urge the interested reader to look at the philosophy, content, and experiences of the other projects.

The CCH Project: Philosophy and Course

The Calculus Consortium based at Harvard is a consortium started by the following individuals and institutions:

Deborah Hughes Hallett, Harvard University
Andrew Gleason, Harvard University
Sheldon P. Gordon, Suffolk Community College
David Lomen, University of Arizona
David Lovelock, University of Arizona
William McCallum, University of Arizona
Brad Osgood, Stanford University
Andrew Pasquale, Chelmsford High School
Jeff Tecosky-Feldman, Haverford College
Joe Thrash, University of Southern Mississippi
Karen Thrash, University of Southern Mississippi
Tom Tucker, Colgate University

In our early discussions and planning, we came to believe that the primary problem with the traditional calculus approach was that it focused almost exclusively on symbolic manipulation; the power of symbolism in mathematics is so compelling that it has tended to force out other approaches to the subject. However, understanding of the mathematical concepts is often better conveyed by geometric images and by numeric approaches. We therefore sought to achieve an appropriate balance among the three approaches. The underlying philosophy behind the CCH project has come to be known as the *Rule of Three*: every topic in calculus should be approached geometrically, numerically, and symbolically. In addition, we have since realized that much of the mathematics is also conveyed verbally, by language, so that our approach could be called the Rule of Four instead.

In order to implement the Rule of Three (or Four), we found it was necessary to redesign both the content and the overall focus of introductory calculus. The result is a very different calculus experience, one that focuses heavily on developing mathematical thinking on the part of the students and less on developing manipulative skill. An old adage says: "You take calculus to learn algebra;" we hope to replace that with a new adage: "You take calculus to learn calculus."

The key to achieving such a goal is not merely to give a new presentation of the material. For most students, what is important is what we expect them to do: the homework and exam problems. Thus, an important part of our work has been to create problems that reflect the Rule of Three, challenging questions that require the students to think mathematically, to understand and to apply the concepts of calculus. For most students, this is something that initially seems strange and demanding: they have seldom been asked to think about mathematics, only to perform rote manipulations that mirror examples in the text.

Historically, the significance of calculus has been linked to its power at solving important problems, typically in the physical sciences. The applications of calculus make the subject important to the overwhelming majority of students, and so we use this to motivate the development of most of the mathematical concepts. Our course is highly problem-driven. Over the last few decades, however, the applications of calculus have grown to encompass areas such as probability, biology, economics and finance. We have consequently included such applications in our materials as well as the usual applications from the physical sciences and engineering.

In many people's view, the primary purpose of taking calculus is to prepare for a subsequent course in differential equations. As one immediate result of our focus on the applications of calculus, we have incorporated a major chapter on differential equations. Our belief is that this topic should be an integral part of calculus, not just the "next course." We consider differential equations as an opportunity to present ideas on mathematical modeling and throughout look at applications from the physical sciences, the biological sciences and the social sciences.

We have also taken the opportunity to present the material from the point of view of what is important in mathematics today. For example, we emphasize the ideas of local versus global behavior of a function. We focus on how the tangent line represents the best linear approximation to a function at a point. We extend the treatment of optimization to consider analyzing the behavior of families of functions with one or two parameters. We view the behavior of the solutions of differential equations using the slope or tangent field.

The Role of Technology

Most of the impetus for calculus reform can be attributed to two factors:

- the growing need for a better educated and more mathematically literate people so that this nation can compete effectively in the international marketplace. As we mentioned above, the traditional calculus courses have simply been ineffective and unsuccessful in preparing and motivating students for technically oriented careers.
- the growing availability of sophisticated technology. For instance, Lynn Steen reported in 1987 on the results of a survey he had conducted of calculus final exams from all types of institutions: 90% of all questions could be answered using widely available computer algebra systems [2]. His conclusion was we should not be teaching to machines, but to people. Since then, cheap and powerful graphing calculators are rapidly becoming a mainstay of mathematical education. Their existence forces us to reassess what is important for our students to be able to do and what is important for them to know.

Our response to this issue was to set a "technology floor," the minimum level of technology that we presumed would be available to all students taking the course. This consists of either possession of a graphing calculator or access to a computer graphics package that will:

1. graph a function;
2. locate the zeros of a function either geometrically or numerically;
3. perform numerical integration;
4. display the slope field associated with a differential equation.

However, we made a firm decision that we would focus on the mathematical ideas, not on the technology which would be present primarily in the service of the mathematics.

Our materials have since been used in a wide variety of technological environments. Perhaps most striking are the experiences at one of the Consortium schools, the University of Arizona, during early class testing of the materials. During one semester, seven sections were taught from our materials: two of these were taught in computer laboratories where each student was sitting at a computer, three with only graphing calculators, and three violated our "floor" and taught with nothing more than an ordinary scientific calculator. The results were a set of totally different courses based on the same text materials; yet each of the instructors was thrilled with the results of his own course.

Clearly, technology has a role to play in calculus, and in all the courses that lead to calculus, as well as those that follow calculus. However, it need not be a dominant role provided the focus is on the mathematics, and the mathematics presented takes into account what is truly important in terms of what technology is able to provide.

Status of the CCH Project

The CCH materials [1] are currently (academic year 1993–94) being used at over 350 institutions in this country and several abroad. The types of schools using it include highly select colleges, large state universities, engineering institutions, small four-year schools, two-year schools, and some high schools.

Student Reactions

The student reaction to calculus reform has been somewhat surprising. Most students seem excited and stimulated by the approach. The emphasis on conceptual understanding and mathematical thinking provides a very different perspective on what mathematics is all about, even if it is intellectually challenging. A constant refrain we hear is "this is the first time I have ever understood mathematics." Such a response is particularly common with weaker students, or at least those with relatively poor algebraic skills that we automatically consider to be weaker students. By providing them with visual and numeric approaches to the mathematics, we are giving them alternative routes to mastery of the concepts of calculus. They thrive. On the other hand, students who have previously aced calculus in high school and who are taking it again in college to get an easy A tend to complain more, at least at the beginning. We hear "This isn't calculus. When are you going to show us that x^7 is equal to $7x^6$?" or "The old calculus was much easier. You didn't have to understand what you were doing to get the right answers."

We have observed some extremely positive outcomes. Students have become far more involved in the mathematics; they care about the answers to the problems because the problems mean something to them. It is no longer just a matter of getting something to match an expression in the back of the book. As a result, they typically work more at the problems.

The students also seem to develop a much better appreciation for mathematics and its importance. They see that mathematics is more than just manipulating quantities; rather it gives them the tools for solving significant problems in different areas. We have received reports from many of the schools that a surprising number of students come out of this course opting to major in mathematics.

We have also received numerous reports of better results in the course. Many schools indicate higher success rates in the sense that there are fewer F's, D's or withdrawls. The alternate routes that the Rule of Three provides certainly seems to help many of the students. At the other extreme, some instructors have indicated that they are giving fewer A's. They suggest that it is very hard for students to ace an exam featuring a series of conceptual problems that require deep understanding and thought compared to the potential ease of doing well on traditional exams that test manipulative skills.

Another characteristic we have observed is a higher level of *persistence* on the part of students. In traditional courses where the emphasis is on techniques, students with weak algebra backgrounds quickly become lost and drop early. The course has only reinforced their negative views both of themselves and of mathematics. We often see a very different attitude in the students taking our course. They tend to feel that calculus is accessible to them and that they are getting something valuable out of the experience, even when they are not doing terribly well. Consequently, they stick with the course instead of dropping out early.

We note that most of the reports mentioned are purely anecdotal. We are now planning some systematic evaluation to obtain more formal conclusions.

Faculty Reactions

Most faculty teaching our course have been extremely positive about their experiences. Most report a sense of personal excitement in teaching *mathematics,* not just *algebra.* In retrospect, many now feel that they used to spend 80% to 90% of their time in calculus doing algebra at the board—either giving examples of techniques or going over problems to find where the students made some algebraic error.

They also report a very different classroom dynamic, with less of the pure lecture mode and more of an interactive educational environment. This might include more discussion between instructor and students; it might include various types of collaborative learning experiences for the students. In large measure, this environment seems to be an outgrowth of the problems in the materials. Since so many of the problems are not routine, many instructors have had to change the way that they deal with homework. Some assign the problems but warn their students that they may find them very challenging and expect the students just to give them their best try. These instructors then use the problems in class the next day as springboards for mathematical discussions. In this way, the students become active participants in developing the mathematical ideas.

Other instructors have reacted to the challenging nature of the problems by having their students work in class in small groups. In some instances, this can be the format of an entire class, in others, an activity used the last 15 minutes of each class hour. In either case, the instructor becomes more of a coach than a lecturer.

Possibly the most telling point raised by many of the instructors teaching the course, including active research mathematians, is the amount of mathematics they personally have learned. We know this is certainly true for most of us who collaborated on the actual development of the materials. Thus, we find that inadvertantly we are having a dramatic impact in the area of faculty development and renewal.

Of course, the fact that some of the ideas presented are new to many users of the materials also presents us with some challenges. We cannot expect that all faculty will be able to pick up the book and be able to use it in class entirely on their own. Rather, many people clearly need training workshops to acquaint them with the philosophy behind the course, the reasons for some of the decisions we made to emphasize or deemphasize particular topics, and to expose them to some of the mathematical ideas we have incorpoated. For instance, we have found that many people teaching calculus are not familiar with slope fields. We have responded to these challenges by presenting numerous training workshops in conjunction with national and regional meetings of MAA and AMATYC, for example, as well as annual workshops at Harvard. We are now encouraging the people who have served as class testers of the materials to provide similar workshops around the country.

In addition, our project has spawned exciting offshoots, local and regional projects designed to expand the implementation and dissemination of our course. Similar activities have grown out of some of the other calculus reform projects. What seems to be particularly effective are local consortia involving universities, four-year schools and two-year schools. Such joint activities usually begin with joint training workshops, not only for faculty teaching the course, but also for the people who provide critical support to the course, such as graduate TAs, student tutors, graders, learning center personnel and part time faculty. Joint activities typically include on-going meetings or e-mail networks; they may involve faculty exchanges and other collaborative efforts to include more institutions, including secondary schools.

Implications for Precalculus Reform

We feel that many of the experiences and ideas we have discussed above regarding calculus reform in general and our project in particular have direct bearing on efforts to reform how students are prepared for calculus. As with the need for calculus reform, there are two primary impetuses for reforming precalculus: a national need for more mathematically and technically trained people and the availability of technology which is changing what is important in the mathematics curriculum.

In developing reform precalculus courses, we feel that our model based on a large group of people representing a variety of institutions has been very successful. We had people from varying institutional backgrounds contributing materials and experiences with different types of students, and we had the advantage of a large base of sites throughout the country at which to test our materials from the outset. We also had the advantage of individuals naturally assuming different, but vital, roles in the project; some focused primarily on writing materials, some on producing problems, some on critiquing drafts, and some on dissemination activities. The broad array of interests and contacts was particularly valuable. Of course, we also had to face the problem that it is more difficult keeping a large group of people in contact with one another, though this can be helped considerably with e-mail. (A small, compact group at a single school or group of neighboring schools has the advantage of being able to meet regularly and inspire one another to work more constantly.) It is also more difficult to keep each member of a large group informed of all developments in a project and to involve each in the day-to-day decisions that have to be made. In all, though, we do feel that a larger group is definitely desirable, particularly in reforming precalculus where the efforts must involve both secondary school and college faculty.

Above and beyond such administrative issues, the key concern in precalculus reform is having a vision for how the course should develop. This may be based on the implications of technology, the learning environment in the classroom, the mathematical content of the course, or the applications that drive the mathematics. Our view is that all of these should be central to developing a better course.

As with traditional calculus courses, precalculus courses tend to focus almost exclusively on algebraic methods. Often, it seems that the course is little more than a semester-long exercise in factoring polynomials and manipulating trigonometric functions to prove endless lists of identities. (Of course, when that course is not particularly successful, many schools simply extend it to a year-long exercise.)

Admittedly, most precalculus courses do focus on graphical ideas. However, as with traditional calculus, the objective is too often on *producing* the graph of a function, whether it is polynomial, rational, exponential or trigonometric. Now that graphing calculators can be in the hands of every student, though, the emphasis should change to focus on mathematical questions—why does the graph appear the way it does? what is the effect of scale on what is seen? what are the local and global characteristics of the function?

Appropriate use of technology should be incorporated into the course. We need to decide on a floor level for technology, such as a graphing calculator. The focus of the course, though, should not be on the technology, but rather on the mathematics we want students to understand and be able to apply. The graphing calculator or other technology should be a tool used in the service of teaching and learning the mathematics.

Moreover, most students taking a traditional precalculus course do not become excited about the mathematics itself, particularly when the emphasis is on mechanical manipulations or even on analyzing the behavior of functions. They want to see some use for that mathematics. Consequently, precalculus courses should be problem-driven. Each topic should be presented in the context of an interesting, motivating example and should be developed in conjunction with a variety of applications. The students should see an immediate tie-in between the mathematics and the world around them.

Precalculus mathematics should focus on mathematical ideas and mathematical thinking as well as on building algebraic skills. Students should be expected to do more than simply reproduce examples given in the textbook. We certainly should expect somewhat less of them at this level than we would expect of students in calculus; but somewhere we must begin expecting something more of them, and the earlier we do, the better it will be for them and for the entire mathematics curriculum.

The topics presented in such a course should be carefully evaluated to determine what is essential, particularly in light of available technology and the changes in the calculus curriculum, and what has become outmoded. For example, in our calculus project, we decided that of the six trigonometric functions standardly studied, only three are truly important; the other three, the secant, cosecant and cotangent, exist primarily for historical reasons to reduce computational drudgery. Look at any sophisicated scientific calculator today—it will have keys for a host of functions that most people may never use, but it does not have a key for these three trig functions. Of course, eliminating these functions does present some challenges, particularly to the instructor, such as having to learn some new identities:

$$\tan^2 x + 1 = \frac{1}{\cos^2 x}$$

or

$$\frac{d}{dx}(\tan x) = \frac{1}{\cos^2 x}.$$

However, we assure you that it is not a problem for students coming upon these relationships for the first time.

The more dramatically that precalculus courses change, the greater the need for faculty training workshops. These may entail learning the use of appropriate technology and will likely also involve learning some new mathematics. They will certainly involve developing an understanding of the philosophy, the rationale, and the implications of the changes that take place in the curriculum. Based on conversations we have had with high school teachers, we suspect that workshops will be particularly valuable at the secondary level.

We also see the collaborative efforts that are developing among local and regional consortia involved in calculus reform expanding to include precalculus (as well as postcalculus) reform. These efforts provide an ideal mechanism for much of the faculty training that will be needed, they provide the mechanism for bringing in the secondary teachers as project participants and for training courses, and they provide the opportunity to develop important and valuable linkages among all levels of mathematics education so that different groups are not functioning in a vacuum.

We do believe that all of this will come about. The calculus reform movement is extremely successful; it can no longer be considered as an experiment. Most people who have used our materials have indicated that they find it inconceivable to go back to traditional calculus. The same comment arises with users of the materials developed by the other calculus reform projects. The face of calculus is changing and will continue to change.

The common thread that now runs through conversations with all the people teaching reform calculus is that there must be a better and different precalculus preparation for their incoming students. Precalculus reform will take place.

References

1. Hughes-Hallett, Deborah, Andrew Gleason, et. al., *Calculus,* John Wiley & Sons, 1994.

2. Steen, Lynn Arthur, "Calculus Today," in *Calculus for a New Century: A Pump, Not a Filter,* MAA Notes 8, Mathematical Association of America, 1987.

A Case Study of a Partnership of Equals: Calculus Meets Precalculus

Arthur Knoebel, Douglas S. Kurtz, and David Pengelley
New Mexico State University

The Calculus Projects Program

A few faculty members at our university began using novel writing assignments in calculus classes in the fall of 1987 and created the name 'student research project' for a two-week assignment involving problem solving and writing. The original projects were a reaction to two things: our students were not learning and performing up to the standards we set; and, since most of our students do not major in mathematics, we found that students would often give their calculus courses low priority. The way we chose to address these issues was simple, on the surface. We decided to give students harder problems to solve and to give them two weeks to solve them. These multistep problems get students to think for themselves, build self-confidence, and alter their view of what mathematics is all about. Students must decide what the problem is about and what tools from calculus they will use to solve it, find a strategy for its solution, and present their findings in a written report. This approach yields an amazing level of sincere questioning, energetic research, dogged persistence, and conscientious communication. Thus, we started to challenge undergraduates in the same way we challenge graduate students; we called these assignments research projects because we viewed them as analogous to our own research.

Within a year, we had been awarded a grant from the National Science Foundation (NSF) for calculus curriculum development. The goals were to determine the feasibility of using projects, to develop a collection of viable projects, and to devise efficient ways to incorporate them into the curriculum. (See [1], [2].) We began with a narrowly focused idea of what we hoped to accomplish: create and use projects. However, incorporating student research projects into traditional calculus classes led us to many other educational issues.

For instance, our students had been trained that writing has no role in a mathematics class. It is a startling realization for mathematicians that this is the logical outcome of contemporary secondary education. We needed to convince students to write mathematics and then train them in scientific writing; in the process we, as instructors, became more articulate in class. As another example, we were led to group work. For practical and pedagogical reasons, we first allowed, then encouraged, and finally expected our students to work in groups.

Our participation in the NSF's initiative for calculus curriculum development kept our attention focused on how to help students learn calculus. But it was clear to us that while the projects we wrote were on topics from calculus, projects could just as well be written for any mathematics course. Though we did not realize it at the time, not only was this pedagogy applicable to other college courses, it was adaptable to precollege instruction. We were soon to begin a collaboration with high school teachers to adapt this method to mathematics classes in the high schools of Las Cruces, New Mexico.

The High School Program

We had spent two years convincing our university students that it was appropriate for them to write mathematical prose, and for this written work to have a major influence on their grades. We came to realize that students should write in their high school mathematics courses as well, both as a better way to learn mathematics and as better training for their future college courses. We then saw that there were no intellectual barriers to high school students tackling student research projects. The gains from having them do so could be significant.

Being research mathematicians, we had little training in educational theory or experience in educational research. We learned much from working on university curriculum development, but when we began to think about incorporating projects into high school classes, we were groping in the dark again.

We began with two premises: high school students could solve projects written at an appropriate level; and we knew neither what that level was nor how to implement projects successfully in a secondary school setting. Several of the things that made our calculus program work—for example, help labs staffed by graduate students, and students working together outside of class—would not work or be available in a high school setting. This made us wary of proceeding. Fortuitously, at the same time we were contemplating this new program, a group of mathematics teachers from the local high schools approached our department asking us for guidance in their courses. We decided to proceed.

During the university curriculum program, we developed a modus operandi that worked well for us. We saw our activities as curriculum development; thus experimentation and modification were encouraged. We also held periodic workshops, about two or three a semester, for all participants to get together and talk. With these experiences in mind, we began to design a high school program.

The four-year program we developed for the high schools comprised one year of small scale experimentation, two years of large scale implementation and more experimentation, and a final year for evaluation. Over thirty teachers have taken part, along with two educational consultants and a statistical consultant. Teachers have been trained in the use of projects and enhanced in problem solving. We explained what projects are and helped them write their first ones. Now they create their own projects, try them in the classroom, and modify the way they are used. Our input was the pedagogy; their input was the implementation. The output will be students excited by the agony and the ecstasy of mathematical discovery.

Our high school program received funding from the NSF. This had two immediate consequences. First, the guidelines in the Teacher Preparation and Enhancement Program of the NSF dictated a distinct hierarchy of personnel. There were principal investigators on the grant, who were the university faculty in charge. There were two levels of teachers: lead teachers, the ones who participated in the program the first year and became a sort of executive committee in later years; and regular teachers, who came in later and were trained by the lead teachers. However, we maintained the philosophy of treating the teachers as our equals, in so far as possible. Second, we were asked to focus our activities on courses that prepare students for science and engineering programs, such as calculus and precalculus. Nevertheless, we initially used projects in algebra, geometry, and trigonometry, and then added calculus in our second year. Projects were appearing in courses lower than the ones commonly designated 'precalculus.' Thus began a fruitful partnership of equals between university faculty and high school teachers.

Further High School Developments

There is a clear and unforgettable sequence any instructor goes through when beginning to assign projects: excitement, about trying something new; fear, upon making the initial assignment; and growth, from incorporating new ideas in the classroom. We saw this happen to college instructors and again to high school teachers. Once the teachers felt some control over what they were doing, they became comfortable with trying ideas on their own.

For instance, many teachers noticed that students were handing in work inferior to the teachers' expectations and the students' abilities. Teachers started to require rough drafts and saw the work improve significantly. (We later employed this idea in our first semester calculus classes.) Grading rough drafts and final drafts became burdensome, but the use of rough drafts was too important to stop. To cut down on time spent grading, several teachers used peer editing in their classes, and some even used it by swapping rough drafts between students in different classes.

As another example, we originally planned to train high school teachers to use cooperative learning activities in their classes. For a year and a half this met with little success, until one teacher came back from a conference enamored with the idea. She helped the group to bring cooperative learning into their classes. These vignettes illustrate some of the difficulties and successes in modifying projects for high schools.

Learning the High School Culture

Originally we thought of the teachers as our equals, but it became very clear, very quickly, that the majority of the teachers would defer to our opinions almost always, even on matters they clearly understood better than we, such as fitting projects into a typical high school class. This was flattering and exciting: it's easy to get hooked on power. We believe that we were fortunate not to get sucked into that trap.

For several reasons, we knew that it was important for teachers to write their own projects. In addition to the obvious reason that they knew the material and the students better than we did, projects written by the teachers get them personally and emotionally involved in the success of their students.

We realized that the teachers needed to have a stake in the program, and we felt we could best accomplish that by allowing them to make decisions as the program evolved, and even, at times, compelling them to do so. We saw our role as guides and facilitators, and whenever major decisions were made, we involved the teachers in them. At times, this was done in a subversive way. For example, a large part of each workshop consisted of a debriefing session where the teachers related their experiences since our last meeting. We took notes on these discussions and wrote all the participants a summary of the day's activities. We edited these notes to include the ideas we thought were good, and to exclude the ones we did not like. In this way we directed the activities the teachers tried over the next several months. Even so, all of the ideas we reported in these notes were initiated by the teachers. Although it was clear that we were leaders, we ran the program as a partnership in so far as possible.

The teachers learned, after we had worked together for a while, that we took them and their opinions seriously. This was accomplished by listening carefully at the workshops and by attending their classes every week. By observing them frequently, we made them feel more important and capable. As we mentioned above, the teachers actually approached us before we approached them. They were ready to change. But they did not have either the knowledge, the support, or the courage to do it. One of the best things we did for them was give them confidence.

To be sure, we do not agree with everything they do. Some matters we fought over and convinced them they were wrong. Others related to the realities of high school classes that college faculty do not have to deal with; they taught us we were wrong. As a result, we do not believe that meaningful, long term change will come about by university faculty prescribing the changes. It must make sense to the teachers and fit into their world, and it is impossible for us to see how to do that from our vantage point.

It pleases us to think that we have helped empower the teachers to change. They are evolving as individuals and as a group. In addition to the initiatives they introduced into the program, they are embarking on even bolder things. Several teachers are making presentations on the use of projects at local and regional meetings of the National Council of Teachers of Mathematics (NCTM), and at an annual national joint meeting of the MAA and AMS. Teachers have designed, written, and received funding from the state for two proposals, one to train middle school teachers to use projects and another to create additional projects for high school students.

The Mathematics Learning Center

Precalculus courses at New Mexico State University are taught in the Mathematics Learning Center. Our department created this center to address the needs of a very large and diverse group of students that arrives at the university with deficient mathematical skills. Since many of these students go on to our calculus program, we would like them to have preparatory work on projects and writing assignments in Learning Center classes. This presents new challenges since the Learning Center was structured to make efficient use of resources by providing standardization of course material, uniform performance expectations, and a focus on skill development. In fact, ten full-time teaching faculty, two part time instructors, fifteen graduate assistants, and many undergraduate tutors taught over 4,000 students in the Learning Center in 1992. The Learning Center originally taught only self-paced mastery-based courses. Eventually, lecture courses were introduced, but these courses are graded by the same mastery standards.

This lean structure presents obstacles to introducing projects into Learning Center classes. The Learning Center's teaching faculty have little time available for curriculum development. Also, students in a single class are at many different places in the syllabus at any one time. Still, we felt that projects would challenge and benefit these students. We again faced the problem of how to implement projects in a new framework.

This past fall, the director of the Learning Center proposed a developmental program to adapt the projects approach. It attempts to allow the unique and successful structure of self-paced mastery classes to continue, and to accommodate the schedules of faculty members who currently have an average load of three lecture and three mastery classes. Some of the novel ideas to be pursued are computer-generated projects and 'grading groups,' that is, faculty and graduate assistants who meet regularly to grade projects.

Teacher Training

An old adage states that a person teaches the way he or she was taught. This points out the biggest and most long lasting way in which we will influence precalculus instruction. Future high school teachers sit in our classes right now, and the way we teach them influences the way they will teach. Students currently taking our calculus courses are learning about projects and about themes, an adaptation of the projects idea that incorporates the learning of core curriculum material into group work and writing assignments. Also, we have used projects in a geometry course for future high school teachers.

Finally, our curricular innovations can work a change in another way that we have not pursued at length. Largely by running summer courses, we have trained local high school teachers to use projects in their classes. We could easily adapt these summer sessions to create a new methods course for students in mathematics education. Just as in our calculus classes where we have students do projects, such a new course might first have future teachers do some projects, then train them in how projects are used, and finally give them practical experience in writing their own materials. Such a course would necessarily address issues concerning cooperative learning and student writing.

Conclusion

It is clear to us that university faculty can and should take part in curricular change at the precalculus level. We have valid ideas about what mathematics is important and how it should be taught. High school teachers and full-time teaching faculty do not always have the time, and often not the confidence, to initiate change. These are important things we can supply. Our experiences lead us to believe that the best opportunity for success and long term implementation lies in forming partnerships in which people with different backgrounds have approximately equal stature and all participants have a stake in the outcome of the program. This was true when making the use of projects a department-wide program in our mathematics department, and no less true when working with the high school teachers or when planning with our Learning Center faculty. Any change we hope to effect must incorporate the realities of their daily lives — they will always have a better understanding of that than we will. Most importantly, ownership makes people strive for success. We believe that our collaboration with high school teachers points to a way for university faculty to take part in precalculus reform.

References

1. Cohen, Marcus, Edward D. Gaughan, Arthur Knoebel, Douglas S. Kurtz, David Pengelley, "Student research projects in the calculus curriculum", chapter in *Priming the Calculus Pump: Innovations and Resources* (MAA Notes, No. 17), Thomas Tucker (ed.), Mathematical Association of America, Washington, D.C., 1990 .

2. Cohen, Marcus, Edward D. Gaughan, Arthur Knoebel, Douglas S. Kurtz, David Pengelley, *Student Research Projects in Calculus* (MAA Spectrum Series), Washington, D.C., Mathematical Association of America, 1991.

A Report on a Project to Develop Course Materials to Integrate Precalculus Review with the First Course in Calculus

Doris Schattschneider, Alicia Sevilla, and Kay Somers
Moravian College

In 1988 the Mathematics Department at Moravian College introduced a two-semester course that intertwines precalculus topics with the material in a first calculus course. This course replaced a traditional two-semester Precalculus-Calculus I and a Precalculus-Applied Calculus sequence. Since the summer of 1991, some members from the Mathematics Departments at Moravian College and Northampton Community College have written materials for the course. This work has been funded by a 2-year grant from the Fund for the Improvement of Post-Secondary Education (FIPSE), U.S. Department of Education.

The Course

The new sequence, introduced in the Fall of 1988, consists of two one-semester courses: Calculus I with Precalculus Review, Part I, and Calculus I with Precalculus Review, Part II. These two courses together cover the same calculus material as a standard Calculus I course, plus review topics in algebra, elementary functions, and problem-solving. These review topics are introduced when needed for the calculus topics and put into the context in which they will be used to solve calculus problems. The calculus topics covered in the first semester course are limits, continuity, and derivatives (including rules of differentiation and implicit differentiation) for algebraic functions. The calculus topics included in the second semester course are derivatives of trigonometric, exponential and logarithmic functions, extreme values, curve sketching, other applications of the derivative, antiderivatives, area and the definite integral, and the Fundamental Theorem of Calculus. The review topics are integrated with these calculus topics throughout the two semesters.

To facilitate the transition to Calculus II for those students who continue the calculus beyond their first year, we use the same calculus textbook that we use for the standard 3-semester sequence, Calculus I, II, III. When we first prepared the new course in 1988, we found that there were no supplemental materials available that were designed with the intention of integrating review of precalculus material in a slower-paced calculus course. For the first three years in which the course was taught, we used different College Algebra review texts to provide the supplemental material but found these texts were not well suited to our purpose.

This led to our proposal to FIPSE to develop suitable materials ourselves. There are three main goals for the supplemental text that we have written:

1. review the precalculus concepts as needed for calculus;

2. motivate the students to study mathematics in general and calculus in particular with applied problems from their sphere of experience;

3. provide exercises and examples that break complex ideas into smaller, more manageable parts.

The Text

Our text, *A Companion to Calculus,* begins with an introduction that describes in general terms what calculus is, the fundamental role of functions in the study of calculus, and the use of symbols in mathematics. It is stressed that the language of mathematics includes four modes: *words, pictures, numbers,* and *symbolic formulations,* and that students need to be able to communicate in these four modes and be able to move from one mode to another. The introduction also contains exercises so that students can practice these transitions. Throughout the text we aim to present each topic in as many of the modes as possible.

The first goal of the materials is to review precalculus concepts as needed for calculus. The basic topics of Cartesian coordinates and functions are covered in Chapters 1 and 2. Different ways to represent functions using the four modes of the language of mathematics are stressed. Simplification of algebraic expressions and the solutions of linear and quadratic equations and inequalities are reviewed in several different chapters. For example, linear and absolute value inequalities are discussed in Chapter 3, *Companion to Limits.* Inequalities that involve quadratic functions are reviewed in Chapter 4, *Companion to Continuous Functions,* and these and more general inequalities are examined again in Chapter 16, *Companion to Extreme Values of a Function.* Methods to find or to approximate zeros of polynomials are also reviewed in several chapters.

Chapter 6, *Rates of Change,* provides a thorough presentation of the concepts of rate and average rate of change of a function before the introduction of the concept of derivative, which is an instantaneous rate of change. In Chapter 9, *Companion to Implicit Differentiation,* examples show how to solve an equation for $\frac{dy}{dx}$.

Rules of exponents for rational exponents are discussed first in Chapter 7, *Companion to Rules of Differentiation and the Chain Rule* and are reviewed and extended to real exponents in Chapter 13, *Companion to Exponential Functions.* Chapter 7 also contains a section on decomposition of functions which prepares students to recognize when it is appropriate to use the various differentiation rules. This topic is addressed again in Chapter 18, *Companion to Antidifferentiation.*

In addition to reviewing precalculus topics as needed for calculus, the materials are designed to reinforce ideas from earlier in the course. The following exercise in Chapter 13, *Companion to Exponential Functions,* relates exponential functions to functions discussed in Chapter 2.

Exercise Match each of the following functions to its graph below.

a. $f(x) = 2^x$ d. $F(x) = x^{1/2}$
b. $g(x) = x^2$ e. $G(x) = e^2$
c. $h(x) = (1/2)^x$ f. $H(x) = \frac{1}{2^x}$

The statement of the problem is followed by six graphs.

A second goal of the materials is to motivate students to study mathematics, and in particular, calculus. In Chapter 1 a map of Washington, DC, is used to motivate a discussion of the use of Cartesian coordinates and the distance formula. In Chapter 2, *Functions,* step functions are illustrated with the following example.

Example 2.11 To make an operator-assisted telephone call to London, the phone company charges $5.50 for the first three minutes and $.75 for any additional minute (or fraction of a minute). Draw a graph and also describe in symbols the function that represents the charges (in dollars) as a function of the length of the call (in minutes). How much does it cost to call for 10 minutes? $8\frac{1}{2}$ minutes?

This example, as with each example in the *Companion,* is followed by a full solution. At the end of each section there are similar exercises for students to work. The following exercise appears in Chapter 2 as a follow-up to the example above.

Exercise The following table indicates the dose of a medication a child should receive according to the child's weight. In applying the rule, all weights are rounded up to the next integer. Thus the dose for the weight 47.3 lbs. is 2 teaspoons.

Weight (in pounds)	Dose (teaspoons)
24–35	1
36–47	1.5
48–59	2
60–71	2.5
72–95	3

Let f be the function that assigns to each weight the corresponding dose.

a. Draw a graph of f.
b. Describe f in symbols using function notation.
c. What is the dose for a child who weighs 46 lbs.?

Here is an example from Chapter 5, *The Role of Infinity.* It illustrates the four modes of communication stressed throughout the materials—words, pictures, numbers, and symbols.

Example 5.1 The potency (in milligrams) of vitamin C tablets is a function of the time t they have been stored. The graph in Figure 5.1 shows this relationship. Explain what the graph shows. Does the graph seem reasonable?

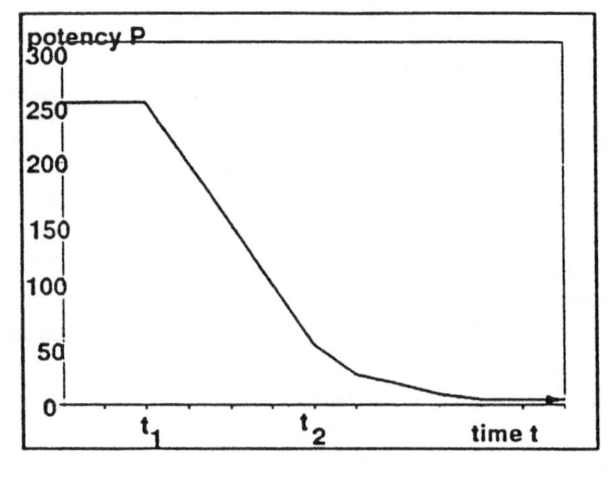

Figure 5.1

Solution: The graph shows that the tablets have an initial potency of 250 milligrams and potency remains constant at 250 milligrams until time $t = t_1$. From time t_1 to time t_2, the potency declines to 50 milligrams in a linear manner. After time t_2, the potency continues to decrease, but at a slower pace than before. The potency is actually never 0, but it is small for large values of t. The larger t is, the closer the value of P is to 0. In other words, storing the vitamins for a long time results in tablets with very little potency.

A third goal of the materials is to provide examples and exercises which break down complex ideas into smaller, more understandable parts. The following examples, the first from Chapter 11, *Companion to Related Rates,* and the second from Chapter 16, *Companion to Extreme Values of a Function,* illustrate how to set up equations in order to solve problems.

Example 11.1 A radio tower has height 130 feet and has been assembled on the ground, lying on its side. A motorized device raises the tower until it is in its vertical position.

a. Draw and label a picture to represent this situation, and identify any quantities that change as time changes.

b. Find a relationship between the variable quantities and express the relationship as an equation.

Example 16.5 A farmer wishes to fence a rectangular field to enclose an area of 9,000 square feet. The south and west sides of the fence will cost \$3.80 per foot and the other two sides will cost \$4.05 per foot. The farmer wants to know what dimensions of the field will minimize his cost. Give the function and its domain that he should consider in order to find the answer. (Assume that the farmer's land is a square, 1 mile on each side.)

Evaluation of the Course and Text

We tested a draft version of *A Companion to Calculus* in all four sections of Calculus I with Review (over 100 students) during the 1991–92 academic year and have used a revised draft in 1992–93 and 1993–94. To assess the effect of this course and our materials, we carried out a four-part evaluation. A two-year evaluation report for the 1991–92 and 1992–93 years is available by request.

First, we used a survey to evaluate student attitudes toward learning mathematics, perception of what was important in the course, opinions about the *Companion* material, the calculus text and the course. This survey was administered to all the students in the course at the end of each semester.

Second, final examinations were designed in cooperation with the instructors of the four sections of the standard one-semester Calculus I. To compare the performance of students in this course with that of students in the one-year Calculus with Review course, 12 of the questions on the standard Calculus I final exam appeared on the final exam of Calculus I with Review Part I and 13 others appeared on the final exam of Calculus I with Review Part II.

Third, to measure the improvement in understanding the review material, the final examination of Calculus with Review Part I also contained questions from the departmental placement test administered to all freshmen before registration for the course.

Fourth, data on course enrollment, attrition, and completion of the Calculus I with Review course were compiled for all years beginning with 1988 and compared with similar data for the Precalculus-Calculus I sequence for the years 1986–1988.

The results of the evaluations gave us some hopeful signs of success. On questions that were common to the final exams in the "regular" Calculus I and the Calculus I with Review, students in the first course did only slightly better than those in the latter. In the Fall exam, evening section (adult) students in the latter course actually outscored those in the "regular" course on 3 (out of 12) questions, and in the Spring exam the mean score for the Calculus I with Review students was significantly higher than that of the Calculus I students on one of the 13 common questions.

The comparison of individual student scores on questions on their placement examinations and scores on the same questions on the final examination for the Calculus I with Review course showed that most improved their scores. A comparison of mean scores for common questions on the two tests showed improvement in every case.

The data on attrition and course completion have also been encouraging. The continuance rate from the first to the second semester has increased. During the two years 1986–87 and 1987–88, only 65% of the students who passed Precalculus the first semester continued to Calculus I or Applied Calculus in the second semester. During the five years of the new course, 1988–93, 80% of the students who passed Calculus I with Review Part I continued to Part II the next semester. The success rate in completing Calculus I in one year has improved. For the two years 1986–87 and 1987–88, 40% of the students who enrolled in Precalculus successfully completed Calculus I or Applied Calculus the same year. During the last five years, 50% of the students who enrolled in Calculus with Review Part I successfully completed Parts I and II the same year.

The results of the student attitude survey have also been encouraging. Most students indicated the *Companion* text was helpful and preferred it to the calculus text. Most indicated an improvement in their ability to solve mathematical problems and felt the material could be useful in their major.

Other favorable outcomes of the course have been noted. Transition into Calculus II has been no problem; in fact, we have had a few students who started in Calculus I with Review and continued to work for a computer science or mathematics major quite successfully. Students who pass Calculus I with Review Part I and choose not to continue to Part II learn something beyond the material they had in high school, unlike when they took precalculus and did not continue on to calculus.

The *Companion* has been especially successful with adult students who have not studied mathematics for a long span of years but are highly motivated. In the 1991–92 year, 28 of the 29 adult students who began Calculus I with Review Part I successfully completed the full-year course.

On June 18–19, 1993, a dissemination conference was held at Moravian College. Sixty participants from 40 institutions showed interest in this integrated approach to teaching calculus to under-prepared students; we have learned of several colleges that have developed their own integrated 1-year course. As a result of dissemination activities, eight institutions are using the draft version of the *Companion* in 1993–94. A commercially published version (by Brooks/Cole) of the text will be available for classroom use in 1994. We welcome class-testing in a variety of environments.

Postscript:

In April 1994, this project was chosen by *Project Kaleidoscope* as a "Program that Works."

Precalculus Papers

Mathematics Now, Leading to Calculus of the Future

Paul A. Foerster
Alamo Heights High School

The following are some things I have done with my students in Trigonometry/Precalculus to prepare them for AB or BC calculus at our own school, or for calculus in college. Annotations in italics tell some of my observations and conclusions based on students working these exercises.

Each student checks out a TI-81 calculator along with textbooks from the school. Thus, a graphing calculator is available to all students at all times, both in class and at home. However, the course is designed to develop mathematical concepts in students' minds, rather than to "teach using the graphing calculator," or to "implement the NCTM *Standards.*" Anything that enhances learning the concept is used, calculator, computer, pencil and paper, or sheer brain power.

Originally, each paper was written in Microsoft Word 5.0 for the Macintosh. The graphs were drawn with PS-MathGraphsII, which allows one to draw highly accurate graphs with many options, save the graphs in encapsulated PostScript, and drop them into a word processor document. The software is available for both Macintosh and IBM.

You are welcome to use the ideas from these papers with your own students. You may even duplicate them for use verbatim, if that suits your purpose. However, it is requested that you do not publish the papers in your own works without first consulting me and without acknowledging their source.

Trig/Precalc

More Harmonic Analysis

Objective: Given a graph that is a sum or product of sinusoids, find its equation.

1. Find the equation of the graph in Figure 1. Confirm your equation by TI-81. Sketch on the graph the two sinusoids that were combined to form it.
2. Find the equation of the graph in Figure 2. Confirm your equation by TI-81. Sketch on the graph the two sinusoids that were combined to form the graph.

[*The problems in this exercise are made accessible by (1) the ability of the instructor to draw "anatomically correct" graphs, drop them electronically into a word processor document, and print them on paper for students to use; and (2) by the ability of students to draw the results on their grapher, thus confirming their answers. Mathematics can be more easily taught in the "Predict, then Do" mode now.*]

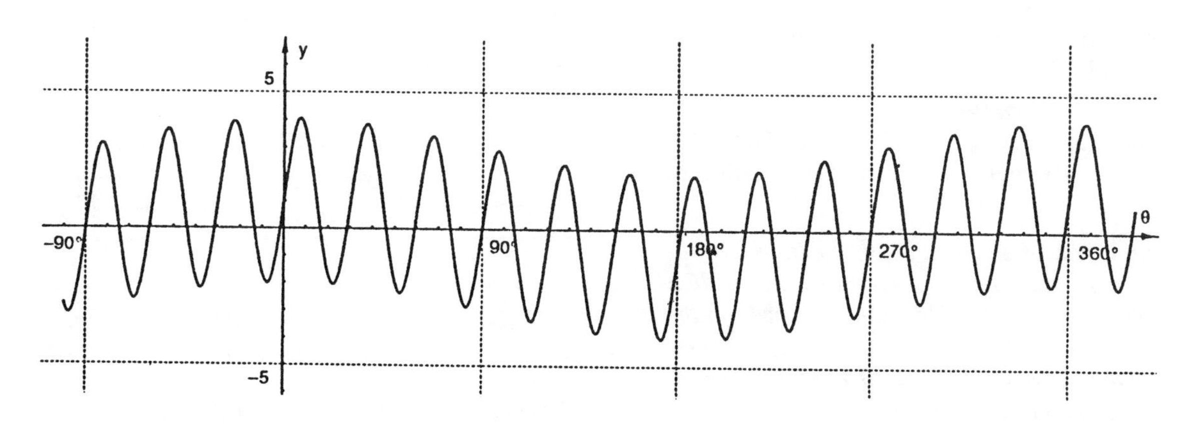

Figure 1:

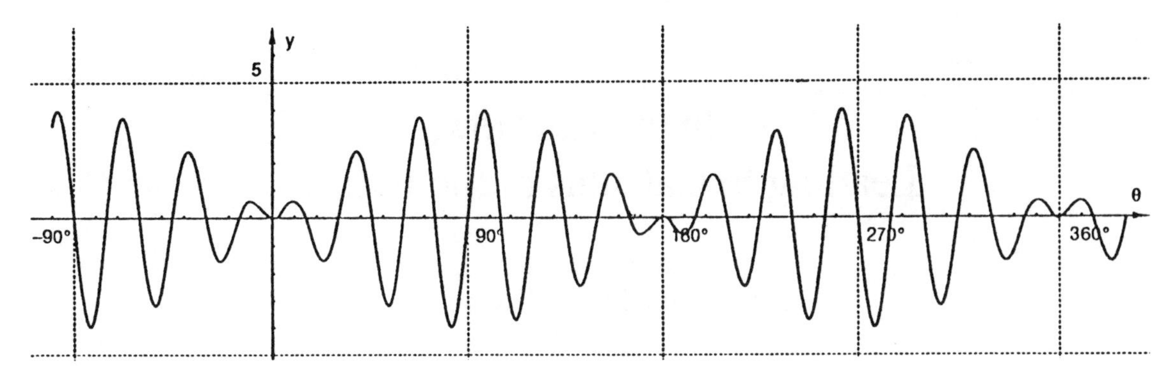

Figure 2:

Trig/Precalc

Variable Triangles

Objective: Show how the third side of a triangle varies as the angle included between the other two sides increases.

1. The diagram shows a triangle with sides 4 cm and 5 cm, and included angle $\theta = 30°$. Do you agree with these measurements? ____

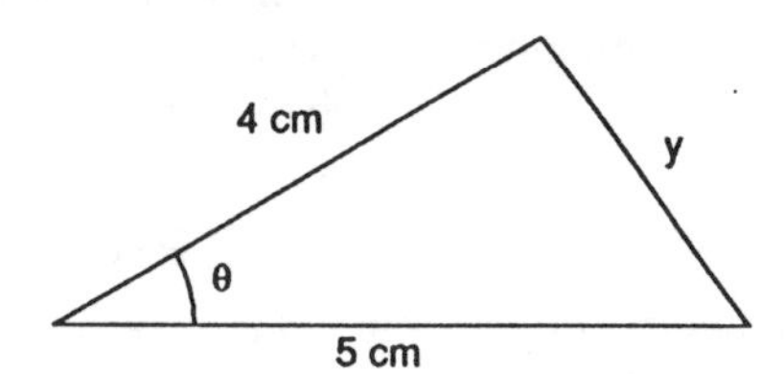

[Since a photocopier can alter the size of an accurately-drawn computer graphic, it is prudent to ask the students if they agree with the measurements. That way, if their computed value doesn't quite match the measured value, they will realize why.]

2. Calculate the length of side y.

3. Measure side y. Does your calculation agree with the measured value?

Measured: ____ Agree? ____

4. Calculate the value of y for each value of θ from 20° through 160°, by increments of 20°. Be as time-efficient with these calculations as possible.

[This problem is part of a continuing theme in which students are led to see that variables really vary, rather than stand for unknown constants. This concept is perhaps the most important one to students preparing for calculus since that subject concerns how fast a variable varies. The introduction to Goodman's calculus texts contain words of wisdom in this regard.]

5. Draw the triangle with $\theta = 140°$ on the diagram in Problem 1. Does the actual length of y agree with the calculated length?

[On the actual work sheet space was provided for this drawing. This problem provides an opportunity for students to study mathematics in a "Predict, then Do" mode. The ability of instructors to draw "anatomically correct" diagrams and drop them electronically into word processor documents allows us to create problems where students measure with a ruler in centimeters. Measurements on paper are more real to students (and less expensive for equipment and time!) than measurements made on a computer screen.]

6. Plot the graph of y versus θ from 0° through 180° on the grid below.

[Degrees are more appropriate than radians here. The "nicer" numbers allow students to concentrate on learning the concept in a more natural setting.]

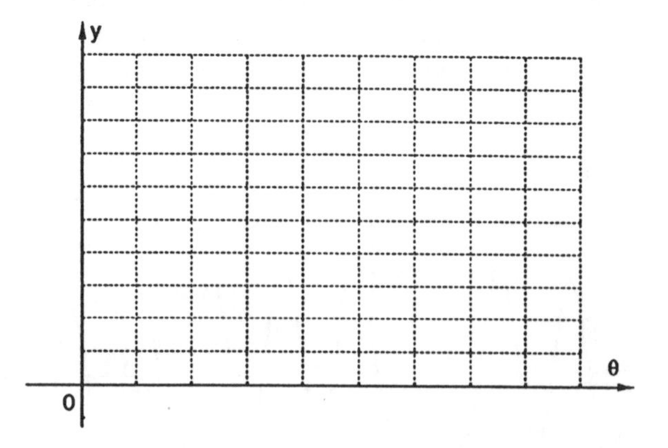

[The grid above was drawn to the desired size by computer software and dropped into the word processor document.]

7. If you have not already done so, put the equation for y as a function of θ into the TI-81. Then plot the graph

of y for values of θ from $-100°$ through $850°$. Sketch the resulting graph below.

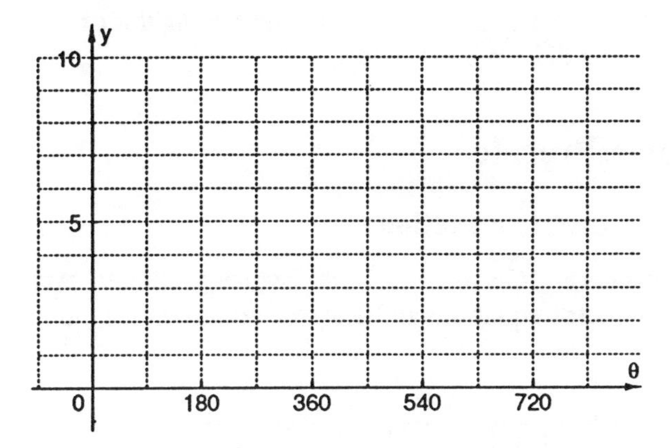

8. The graph in Problem 7 is periodic. Is it also *sinusoidal*? Justify your answer.

[*The graph is not sinusoidal. The ordinates are squares of a sinusoidal function, as you can tell by realizing that y^2 appears in the law of cosines. Exercise like this show students that not all periodic functions are sinusoidal.*]

9. Although triangles cannot have angles that are negative or greater than 180°, the function graphed in Problem 7 *can* represent things in the real world. Write at least one real-world situation to which this graph could apply.

[*I challenge you to think of some!*]

10. What did you learn as a result of doing this exercise that you did not know before?

[*A variation of this question is asked at the end of most exercises. It lets the students tie together what they have learned, and gain experience in writing about mathematics. It also allows the instructor to learn something about what is going on in students' minds, thus allowing a focus on plans for subsequent instruction.*]

Trig/Precalc

The Gaussian Distribution Curve

In this exercise you will plot a "bell-shaped" curve for data that is (roughly) normally distributed. (A normal distribution is also called a Gaussian distribution.)

The table below shows hypothetical data for the lengths of 100 babies born at Scorpion Gulch Hospital. The lengths are grouped into classes. For instance, babies between 17 inches and 18 inches are recorded as 17.5 inches.

Length	Number
13.5	1
14.5	1
15.5	3
16.5	9
17.5	11
18.5	22
19.5	17
20.5	20
21.5	8
22.5	5
23.5	2
24.5	1

1. Find the mean and the standard deviation of the data. Write the results here.

$$\overline{x} = \_\_\_\_ \quad \sigma = \_\_\_\_$$

[*The data here were "rigged" with the help of a spread sheet to give fairly nice values for mean and standard deviation, as well as to give a mode and a median not equal to the mean.*]

2. A bell-shaped curve for these data is given by the equation

$$f(x) = \frac{n}{\sigma\sqrt{2\pi}} e^{-(x-\overline{x})^2/2\sigma^2}$$

where x is the value of the data (inches, in this problem), n is the number of data points, σ is the standard deviation of the data, $\overline{x}$ is the mean of the data, and e is the base of the natural exponential function, 2.71828.... Calculate decimal approximations for the coefficient in front of the e, and for the constant in the exponent. Then write the equation.

[*The number n is included in the Gaussian equation so that the area under the graph will equal the number of data points.*]

3. Plot the graph of the equation in Problem 2. Use an x-range of 10 to 29 (a friendly window!) and a y-range of 0 to 26. The function e^x is built into your calculator.

[*Problems like this give a reason for introducing exponential functions with e as the base. For the time being it is sufficient to say, "Base e must be important because they went to the expense of building it into your calculator!" The familiarity with e makes it less frightening when it appears "for real" in calculus.*]

4. The graph on the TI-81 should look like this:

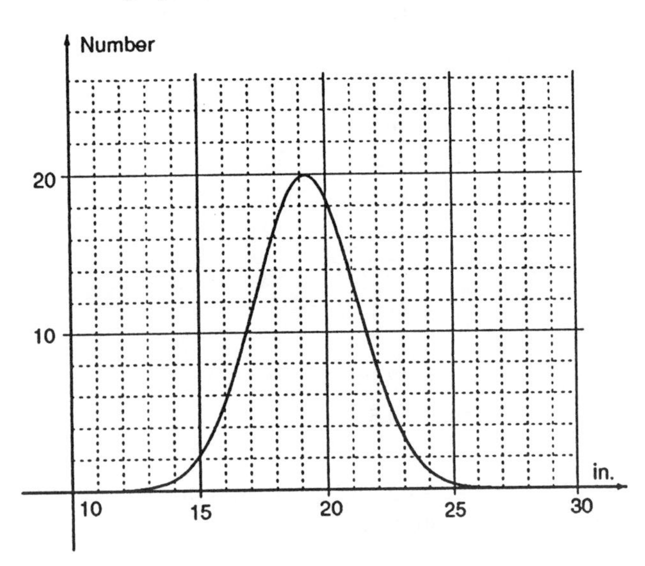

Use the TRACE feature to get values of $f(x)$ for several integer values of x. Do the results agree with the points on the graph? ___

[*Computer graphics allow the instructor to generate "anatomically correct" graphs on paper. The graphing calculator, used appropriately, allows students to confirm that the graph on paper really is the one produced by the equation.*]

5. Find the area of the region under the graph by counting squares. Estimate fractional squares to the nearest 0.1. Record your answer here.

6. How many babies does each square represent? What does the total area represent?

[*Seeing area under a graph representing something tangible, like a number of data points, plants a seed in students' minds that develops later when they study integrals in calculus.*]

7. Approximately what percent of the area lies within ± 1 standard deviation of the mean?

[*Students found clever ways to answer without being told how.*]

8. Plot a histogram of the data in Problem 1 on the graph in Problem 4. How does the area of the bars sticking above the curve compare with the area left out under the curve by bars that don't quite reach?

[*The area under the curve equals the area of the histogram. Thus there is just as much of the histogram sticking up above the curve as there is empty space below it. Here, again, the accurate graph on paper lets students reach conclusions in a cheaper, more time-efficient, and less mysterious way than graphs either on the computer or on the calculator screen.*]

9. What did you learn as a result of doing this exercise that you did not know before?

Trig/Precalc

Parametric Functions

Objective: Given a diagram involving ellipses, write parametric equations and draw the graph on the grapher.

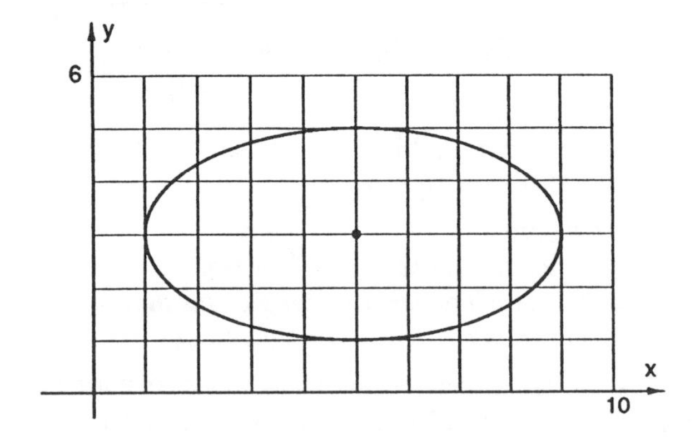

1. The diagram above shows an ellipse with center at $(5, 3)$, x-radius $= 4$, and y-radius $= 2$. Write what you *think* the parametric equations of this ellipse are.

2. Plot these parametric equations on the TI-81. If you use an x-range of $[0, 9.5]$ and a y-range of $[0, 6.3]$ you will have a friendly window. Does the grapher's graph agree with the diagram? ___

[*Students were introduced to parametrics the day before using a pendulum made of a weight from the physics lab hung from the ceiling by string, coming almost to the floor. The pendulum was pulled back 30 cm in the x-direction, then pushed in the y-direction. Students then found parametric equations for the resulting elliptical path. They found out how to enter these equations into their TI-81's, and were amazed to find that the calculator graph matched the pendulum's path.*]

For Problems 3 through 6, write parametric equations and draw the graph on the grapher. You may use the DRAW command to draw straight lines and to put dots in appropriate places.

[*The PSMathGraphsII software I used to generate these graphs allows one to enter equations in parametric form. The ability to draw diagrams by computer graphics is a viable reason for students to study parametric functions. Parametrics can now be studied without first transforming to Cartesian form.*]

3. Cone:

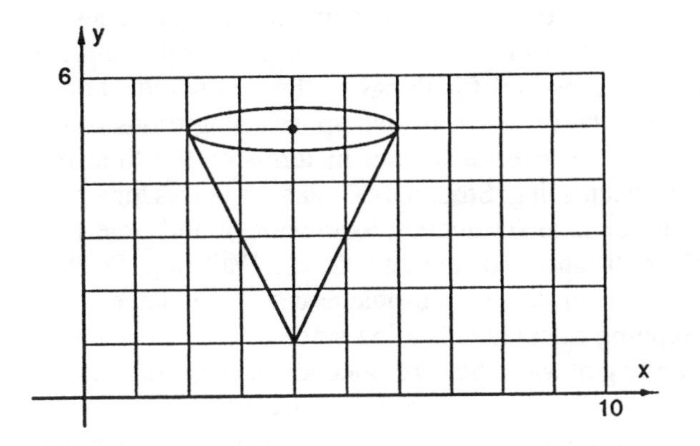

4. Cylinder:

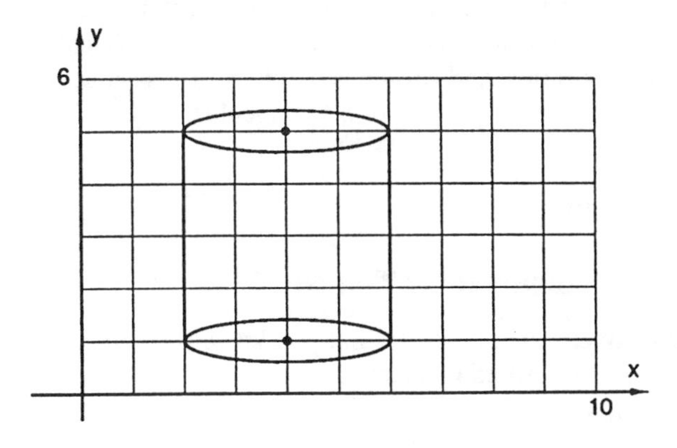

5. Frustum of a cone ("truncated" cone):

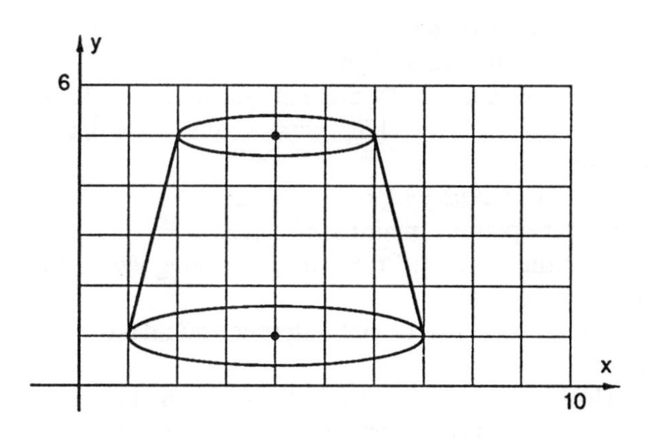

[*Some students figured out how to show only the front half of the lower ellipse by using* $(1/2)t$ *as the variable for that equation.*]

6. Cone oriented the other way.

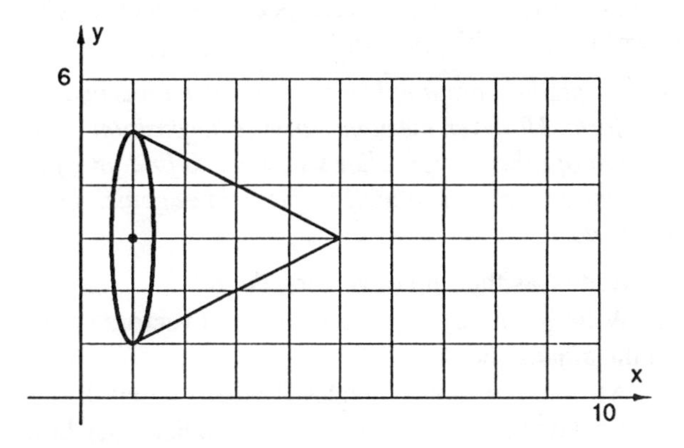

7. What did you learn as a result of doing this exercise that you did not know before?

[*This is an exercise that would be meaningless without the grapher, but is invaluable for conveying the concept of parametric equations in a way that has previously been inaccessible to students.*]

Trig/Precalc

Vector Equation Game: The Planetary Motion Problem

Objective: Use vectors to get parametric equations of the path of a point on a planet.

The diagram shows a small planet 3 miles in radius circling around at 10 miles away from a black hole. Vector $\vec{v}_1$ goes from the center of the black hole to the center of the planet. Vector $\vec{v}_2$ goes from the center of the planet to point $P(x, y)$ on the surface of the planet. An-

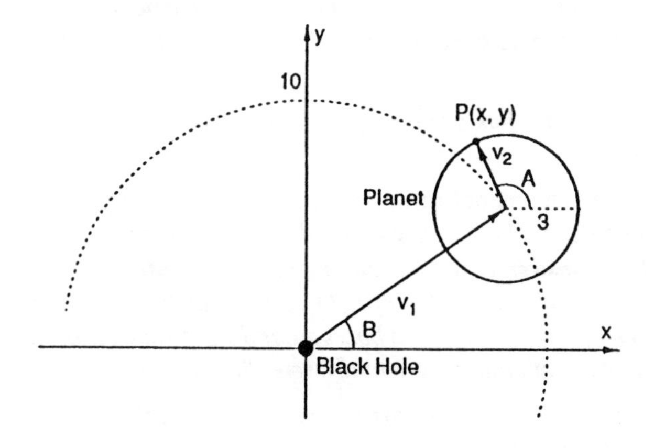

gles A and B are in standard position at the centers of the planet and the black hole, respectively. Vector $\vec{r}$ (not shown) is the position vector to point $P(x, y)$.

[*Vector equations are simply parametric equations in a different form. However, the ability to write the position vector as a sum of other vectors allows students to find an equation of the path without having to spend time agonizing over what to do.*]

1. Write $\vec{r}$ as the sum of vectors, $\vec{v}_1$ and $\vec{v}_2$.

2. Write $\vec{v}_1$ and $\vec{v}_2$ in terms of the unit vectors $\vec{\imath}$ and $\vec{\jmath}$ and the angles and A.

3. At time $t = 0$ both angles A and B equal 0. The planet rotates counterclockwise at 12 radians per hour and it orbits the black hole counterclockwise at 2 radians per hour. Write equations expressing A and B in terms of t.

4. Write the position vector $\vec{r}$ in terms of t and the unit vectors $\vec{\imath}$ and $\vec{\jmath}$. Simplify by combining the coefficients of $\vec{\imath}$ and $\vec{\jmath}$.

6. Plot the path of point P on your TI-81. To get a reasonable window, press ZOOM Standard, ZOOM Square. Pick a t-range large enough to get one complete cycle. You might have to use a smaller Tstep to get a reasonably smooth curve. Sketch the resulting graph on dot paper.

7. Explain why there are only *five* loops in the graph, in spite of the fact that the angular velocity of the planet is *six* times the angular velocity of orbit.

[*The planet makes six revolutions with respect to space, but only five with respect to the black hole since it must rotate more than a complete revolution to make the next loop.*]

For Problems 8 through 10, write the parametric equations here, and sketch the graphs on dot paper.

8. The planet slows from 12 radians per hour to 8 radians per hour.

[*The graph still has loops, but very small, and not as many.*]

9. The planet rotates *clockwise* at 12 radians per hour instead of counterclockwise.

[*The graph has seven loops, pointing outward instead of inward.*]

10. The planet rotates at exactly the right angular velocity to make cusps instead of loops in the path.

[*The answer is $6\frac{2}{3}$ revolutions per hour. Most students guessed 6 or 6.5, which produced what looked like cusps. But upon drawing zoom boxes around a potential cusp, students found that they actually were slightly rounded.*]

11. What did you learn as a result of doing this exercise that you did not know before?

The idea that variables really vary can be brought in long before the course which immediately precedes calculus. Algebra I can involve problems in which students write expressions for things in the real world. For instance, the distances of two approaching trains from a given point can be written in terms of the time they have been going. Students calculate distances for a given value of time (evaluating expressions), and time for a given distance (solving equations). Waits and Demana have shown ways such problems can be animated on the graphing calculator, leading students to a clear understanding of what they are calculating. Algebra II can be presented as a study of various classes of functions, giving students a familiarity with the behavior of exponentials, polynomials, rationals, variations, and irrationals before these functions are encountered in calculus.

The following exercise shows how I have been teaching geometry in such a way that students get a hands-on feeling for measurement, prediction, conjecture, calculator, trig ratios, and variables that vary. These concepts are crucial to the preparation for calculus.

Geometry

Short Rehearsal for Test No. 18

Objective: Demonstrate that you know the properties of angles inscribed in a circle.

Sketch:

1. An angle inscribed in a minor arc.

2. An inscribed angle that intercepts a minor arc.

[*Students have great trouble at first in distinguishing between "inscribed in ..." and "intercepting ...."*]

3. An inscribed angle that intercepts a major arc,and the corresponding central angle.

4. If you were to draw a circle of radius 50 ft, and construct a central angle of 120°, what would be the degree measure of:

The corresponding minor arc? ____
The corresponding major arc? ____

5. For the circle in Problem 4, how long would the circumference be?

6. What fraction of the circumference would the major arc be?

7. How many feet long would the major arc be?

8. The circle has quadrilateral $ABCD$ inscribed in it. Find the degree measure of

Arc AB ___ Arc BC ___
Arc CD ___ Arc DA ___
Angle A ___ Angle B ___
Angle C ___ Angle D ___

[*At the beginning of the course some students reveal that they have no concept of angle. When asked to measure an angle, they pick up a ruler!*]

9. Make a conjecture about measures of the *opposite* angles of a quadrilateral inscribed in a circle.

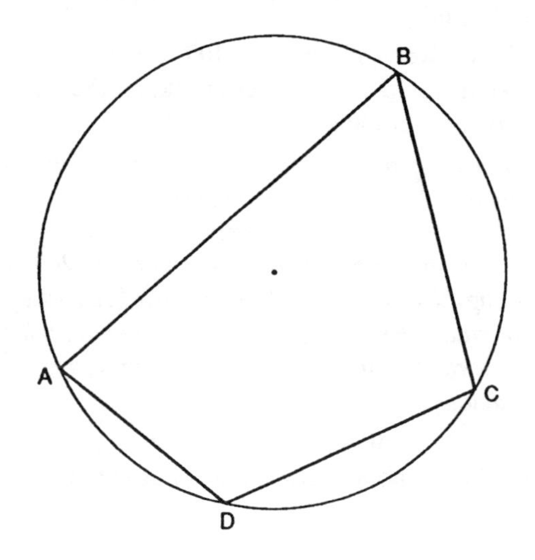

[*The circle, accurately-drawn, allows students to make measurements that lead to a conjecture.*]

10. On the circle below, construct arcs of 30°, 60°, 90°, 120°, 150°, and 180°. Each arc should start at A and go clockwise toward B.

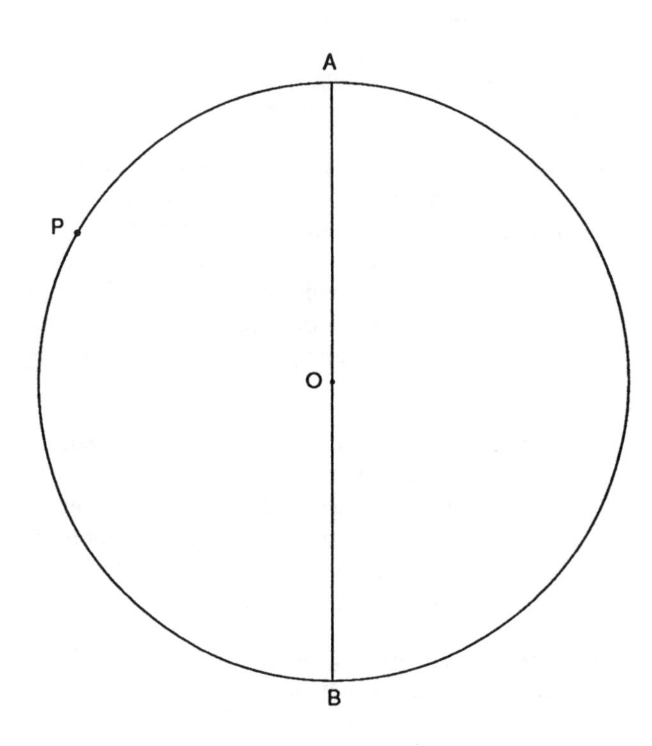

11. Construct inscribed angles with vertex at P that intercept each of the arcs in Problem 10. Measure each inscribed angle.

___ ___ ___ ___ ___ ___

[*Students are amazed to find that the inscribed angles really do come out equal. Once they find inductively what the property says, they are much more willing to learn its proof.*]

12. State the property that explains the relationship between the angle measures in Problem 11 and the arc measures in Problem 10.

[*Problems 10 through 12 show that even in geometry, variables can vary! This concept is crucial for students to grasp if they are to succeed later with calculus.*]

13. State the property that explains why $\angle APB$ is a *right* angle.

[*The rest of this exercise involved proof of the inscribed angle theorem.*]

The calculus course itself can be made more understandable by appropriate use of calculators and graphers. The following is a typical exercise I have used with my calculus students.

Calculus BC

Series Convergence Game!

Objective: Determine whether or not various series of constants converge, and estimate the limit for those that do converge.

1. The geometric series 100 + 80 + 64 + ⋯ can be shown to converge using the *definition* of convergence of a series. Demonstrate that you understand the definition by using it to prove that the series converges.

[*The students were expected to use* $S_n = t_1\dfrac{1-r^n}{1-r}$ *and show what happens to* r^n*, rather than using the shorter formula which gives the limit directly. The instructor's role was to make sure each group of students realized how this formula connects with the definition of convergence of a series.*]

2. The harmonic series $\sum_{n=1}^{\infty}\frac{1}{n}$ diverges, even though the terms get smaller and approach 0 as a limit. Use an appropriate method to explain *why* the series diverges. A graph might help you do the explaining.

[*Students were expected to draw the terms of the series as an upper sum for the integral of* $1/x\,dx$*, rather than simply applying the integral test. The objective was to make sure students understood the concept behind the integral test, not simply to get an answer.*]

The Maclaurin series for $\cosh x$ is

$$\cosh x = 1 + \frac{1}{2!}x^2 + \frac{1}{4!}x^4 + \frac{1}{6!}x^6 + \cdots$$

If you evaluate cosh 2 using this series you get

$$\cosh 2 = 1 + \frac{4}{2} + \frac{16}{24} + \frac{64}{720} + \cdots$$

3. Write, in fraction form, the fifth term of the series.

4. Write, in decimal form, the first three partial sums of the series.

[*The sums are* $1, 3, 3.66666666\ldots$.]

5. Explain why the partial sums of the series are *increasing* although the terms of the series (after the second term) are *decreasing.*

[*Getting students to understand this concepts is as elusive as nailing jelly to the wall. (Just when you think you've got it, it slips off!) It seems like they must some time or other wake up and say, "This doesn't make sense!" before they finally get the "Aha!"*]

It is desired to find out how close the third partial sum is to the actual value of cosh 2 *without* actually evaluating cosh 2. In the next questions you will find this out by getting an upper bound on the tail.

6. Write the first four terms of the tail of the series, starting at term 4. Leave the answers in decimal form.

7. What is the ratio of (term 5)/(term 4)? Leave the answer as a fraction in simple form.

8. Write the first four terms of a *geometric* series with first term equal to the fourth term of the cosh 2 series, and common ratio the number you found in Problem 7.

9. Show that the first two terms of the geometric series are equal to the first two terms in the tail, but from there on, each term of the geometric series is *greater* than the corresponding term in the tail.

10. Find the number to which the geometric series in Problem 9 converges.

11. Explain why the geometric series in Problem 9 is an *upper bound* for the tail of the cosh 2 series.

12. Explain how you can tell from the work above that the cosh 2 series *converges.*

[*Problems 6 through 12 help students understand the concept behind the comparison test. It also lays the foundation for the ratio test, which is simply a comparison test with a geometric series. All of these tests ultimately go back to the property that a series converges if the sequence of partial sums is increasing and bounded above.*]

13. Evaluate the tail of the series by appropriate use of your calculator, and thus show that the number you calculated in Problem 11 really *is* an upper bound for the tail.

[*"Appropriate use of the calculator" is pressing* cosh 2, *then subtracting* 3.666666..., *the third partial sum. The role of the instructor is to point students in this direction, because many of them think they are supposed to start computing terms of the tail.*]

14. If you were to calculate 100 terms of the cosh 2 series, what would be an upper bound on the tail of the series? Justify your answer.

[*The purpose of this problem is to have students repeat Problems 6 to 12, but in a less structured setting. The ratio is* $0.0000985\ldots$ *and the upper bound is* 2.038×10^{-315}. *It took a student with a TI-85 to evaluate the* 200! *which arises along the way. The work would have been truly frightening for students working alone, but is manageable when students work in groups with input and moral support from the instructor.*]

15. As a result of doing this exercise, what do you understand better about series, and what are you still uncertain about?

Conclusions

- The most important thing students must learn in preparation for calculus is that variables really vary. This concept paves the way for derivatives and their inverses, which involve the rate at which variables vary; and definite integrals, that involve a product (such as rate $\times$ time) in which one of the factors may vary. All courses leading up to calculus can help students form this concept in their minds.

- The ability for instructors to create "anatomically correct" graphs using computer technology, and put these graphs on paper using laser printers and word processors opens up a wide range of activities in which students can predict the behavior of a graph or figure, then confirm their results by actual measurement (with ruler and protractor, or by counting squares).

- Graphing calculators in the hands of the students allow verifying the correctness of the accurate graphs on paper, and allow dynamic visualization of mathematical processes. Students using these tools are able to learn mathematics in a "Predict, then Do" mode that was not possible before.

- Work by students in groups, if properly used, lets students develop ownership in the mathematics. As instructors, we can assign tasks difficult enough that we would never dare to assign them for homework. Students often come up with novel methods of attacking a problem that lead both them and the instructor to new insights into a problem.

- Giving students opportunities to verbalize about mathematics, both orally and in writing, helps them develop concepts in their minds, thus paving the way for calculus courses in which the concepts will be needed, and the verbal skills will be expected.

- To encourage the reading of mathematics, we should write it to enlighten our students. Too often we write it to impress our colleagues.

- The lecture method, once touted as a time-efficient way to convey information, does not work well on students today. We as instructors simply cannot compete for the attention of students who are accustomed to learning things via television. We should not regret or deny this fact. We must simply choose ways to accomplish our task with the "raw materials" we have, not with those we wish we had.

- Just as we cannot deny the nature of our students, we cannot deny the existence of technology. Computers and calculators are here to stay. We must convince any reticent colleagues that we have bitten of the fruit of the tree of knowledge, and can no longer go back to the Garden of Eden which existed "B.C." (Before Calculators).

- We must never lose sight of the fact that we are teaching *mathematics,* not applications. We are responsible for teaching mathematical concepts in such a way that students can make connections to other fields. As a person with a degree and experience in chemical engineering, I fully realize the importance of understanding theory if one is to make intelligent applications.

It is an exciting time to be teaching mathematics! Best wishes as we bring the topic into the mainstream for the 21st Century!

Functioning in the Real World: Models for Precalculus Reform

Sheldon P. Gordon
Suffolk Community College

The reform of the calculus curriculum is well under way with changes in emphasis which reflect geometric and numerical ideas in addition to symbolic manipulations. There is also a greater emphasis on applications, the use of technology, and student projects. These new curricula have been adopted at hundreds of institutions around the country, and almost all the sites have reported success and satisfaction. More institutions are now thinking seriously about changing their calculus offerings. The results of these efforts may very well transform calculus into a *pump, not a filter.*

It is now time to consider how we "fill the tank"; that is, how we get more students into calculus. Each year approximately 600,000 college students take a precalculus/college algebra and trigonometry course; yet only about 15–20% of them ever go on to *start* calculus. (Admittedly, this includes many who take college algebra and trigonometry courses which are required terminal math courses; but we should expect a good course will turn students on to mathematics so that they would continue on to calculus.) Most of those students who do go on to calculus from these traditional "preparatory" courses retain little of the material they were taught and do not complete calculus. This is a dreadful indictment of precalculus courses. They neither motivate the students to go on in mathematics nor adequately prepare them when they do continue.

What is needed is a precalculus experience that extends the common themes in most of the calculus reform projects, one that focuses more on mathematical concepts, one that provides students with an appreciation of the importance of mathematics in a scientifically oriented society, one that gives students the skills and knowledge they will need for subsequent mathematics courses, and one that makes appropriate use of technology.

The Math Modeling/PreCalculus Reform Project, with major funding from NSF, is addressing this problem by designing a new way to *fill the mathematics tank.* Our efforts involve developing, testing and implementing a dramatically different alternative to standard precalculus courses. Our goal is to emphasize the qualitative, geometric and computational aspects of mathematics within a framework of mathematical modeling at a level appropriate to precalculus students. We want to capitalize on the fact that most students are more interested in the applications of mathematics than in the mathematics itself, so that the applications should drive the mathematical development. Thus, all the mathematical knowledge and skills they will need for calculus are introduced, developed, and reinforced while applying mathematics to model and solve interesting and realistic problems. Our hope is that such an approach will excite the students and encourage them to go further with mathematics by showing them some of the payoff that a knowledge of mathematics provides.

The goal of our project is to develop a set of materials and a course based on them that will serve a multiplicity of audiences:

- A one semester course that will lay a different, but very effective, foundation for calculus;
- A one or two semester course that will stand as a contemporary capstone to the mathematics education of students who do not plan to continue on to calculus. We expect that the course will encourage many of these students to change their minds and go on to calculus and other quantitatively related courses.
- A course that will provide the foundation for further courses in discrete mathematics and related offerings.

An outline of the materials developed is shown in Appendix I.

The models developed are based primarily on difference equations, data analysis, probability, and matrix algebra. The focus on difference equations includes applications of first and second order difference equations and systems of first order difference equations. The emphasis is on modeling a variety of situations and relating the solutions to the situations. For example, we develop an assortment of models on growth and decay processes (for populations, diseases, technology, etc.) using both first order equations and systems of equations. Our treatment of second order linear difference equations includes a treatment of simple harmonic motion for a mass on a spring as well as the case of a system with damping. Other applications include projectile motion in the plane, effectiveness of sorting methods in computer science, and other models from physics, biology, economics, sociology and so forth.

As an example, consider the problem of modeling the way that the kidneys eliminate a medication from the bloodstream. Suppose a patient takes 50 mg of a certain drug each day and that the kidneys remove 80% of the drug in the patient's blood during each 24 hour period. Let D_0 represent the initial dosage of 50 mg. Then on the following days,

$$\begin{aligned} D_1 &= .20D_0 + 50 = 60 \text{ mg} \\ D_2 &= .20D_1 + 50 = 62 \text{ mg} \\ D_3 &= .20D_2 + 50 = 62.4 \text{ mg} \end{aligned}$$

and so forth. The successive values form an increasing sequence based on the first order difference equation

$$D_{n+1} = .20D_n + 50.$$

We can graph these values, connect the resulting points with a smooth curve, observe that the shape is concave down and that there is a limiting value (known as the maintenance level for the medication). This asymptote can be estimated visually, approximated numerically, or found in closed form. Later in the course, the difference equation can be solved in closed form, using techniques for solving any first order linear non-homogeneous difference equation to produce the specific solution satisfying the initial condition:

$$D_n = 62.5 - 12.5(.2)^n$$

for all n. At each stage, thought-provoking questions are raised: What happens if the patient takes an overdose (a dose above the maintenance level)? or Is the smooth curve used to connect the successive points reasonable?

The techniques used to solve problems involving difference equations are the direct analogs of the techniques used for differential equations. For example, consider what a student must do to solve a second order linear difference equation: The student must find the roots of an associated characteristic equation (using factoring, the quadratic formula, and possibly numerical methods) to construct the general solution of the homogeneous case. He or she must then construct a trial solution for the nonhomogeneous case which involves recognizing classes of functions, performing extensive algebraic manipulations with indices and exponents, and solving systems of simultaneous linear equations. Imposing initial conditions involves solving further systems of linear equations. Most importantly, though, the student must demonstrate an understanding of what information the solution provides about the behavior of the process being studied, particularly in a geometric or qualitative way.

To illustrate this procedure, consider the second-order linear nonhomogeneous difference equation

$$x_{n+2} + 5x_{n+1} + 6x_n = 21 \cdot 4^n$$

The corresponding homogeneous difference equation is

$$x_{n+2} + 5x_{n+1} + 6x_n = 0,$$

whose associated characteristic equation is the quadratic equation

$$r^2 + 5r + 6 = 0,$$

having roots $r = -2$ and $r = -3$. Therefore, the general solution of the homogeneous equation is

$$x_n = C_1 \cdot (-2)^n + C_2 \cdot (-3)^n,$$

where C_1 and C_2 are any two arbitrary constants.

To find a particular solution of the nonhomogeneous equation, we note that the right-hand side is an exponential function, 4^n. Using a discrete form of the method of undetermined coefficients, we assume that the particular solution is also a multiple of 4^n and therefore try

$$x_n = A \cdot 4^n,$$

so that

$$\begin{aligned} x_{n+1} &= A \cdot 4^{n+1} = 4A \cdot 4^n \qquad \text{and} \\ x_{n+2} &= A \cdot 4^{n+2} = 4^2 \cdot A \cdot 4^n = 16 \cdot A \cdot 4^n. \end{aligned}$$

Substituting these quantities into the original difference equation, we find

$$x_{n+2} + 5x_{n+1} + 6x_n = 42A \cdot 4^n = 21 \cdot 4^n,$$

so that $A = \frac{1}{2}$. The complete solution to the original difference equation is

$$x_n = C_1 \cdot (-2)^n + C_2 \cdot (-3)^n + \frac{1}{2} \cdot 4^n.$$

Since this solution involves two arbitrary constants, C_1 and C_2, we require two initial conditions, x_0 and x_1 to determine a specific solution uniquely. Finally, from the fact that 4^n eventually dominates any multiple of $(-2)^n$ or $(-3)^n$, we conclude that the overall behavior of the solution will be one of exponential growth after some possible initial oscillation. The behavior pattern can easily be verified graphically using a graphing calculator or a computer graphics program which displays the solution. It is also possible to examine the first few terms of the solution directly from iterating the original difference equation based on the given initial values.

Furthermore, second order difference equations such as this often lead to pairs of complex roots for the characteristic equation. In turns, this requires that the student uses complex numbers and DeMoivre's theorem in the kind of setting that we eventually want them applied; this is quite different from standard precalculus courses where DeMoivre's theorem is briefly introduced, but never used again until the middle of a course in differential equations several years later.

As another instance, many of the standard formulas for sums of special terms, such as $\sum n$, $\sum n^2$ and $\sum r^n$ can all be developed as simple difference equations whose solutions can be found directly. Thus, rather than handing the students a set of formulas (with no motivation as to where they come from) and then proving them with a poorly understood induction argument, we provide a framework for solving much more varied problems. Further, because of the emphasis on applications, these formulas are then used repeatedly in the sequel to reinforce them.

Data analysis techniques are also introduced early and used throughout so that students learn how to interpret data values that arise in many different contexts. For instance, ideas on regression analysis are developed, including nonlinear regression as a way of reinforcing notions on the behavior of different classes of functions. The computational drudgery is relegated to a computer or graphing calculator with statistical functions. Further, the focus is on collecting and analyzing real data sets and matching them to appropriate mathematical models and interpreting the results.

We expect the students to be actively involved in the process of transforming the data to linearize it. Initially, we require them to identify the likely behavior pattern(s) from a scatterplot, to perform the appropriate transformations by hand, to obtain the best linear fit to the transformed data using their technological tools, and then to undo the transformation by hand. This typically involves extensive manipulations with exponential and logarithmic functions and their properties. Finally, we interpret the results of each analysis and ask appropriate questions about predicting (interpolation and extrapolation) as well as questions about when a given level is expected to be reached. Eventually, we do relegate most of the transformations to the technology, but not before the students understand what is happening and have had the opportunity to reinforce their mechanical skills.

Another thread which is interwoven throughout the course(s) is the notion of probabilistic ideas in the context of performing random simulations. These include Monte Carlo simulations to estimate the value of π, to estimate the average value of a function on an interval, to estimate the frequency of complex roots of polynomials, to study essentially random processes such as radioactive decay and the spread of diseases, and to develop models to investigate various waiting-time situations.

In particular, we emphasize geometric probability as a vehicle for reinforcing geometric and trigonometric ideas in a new setting. Typically, if a random variable is uniformly distributed, the probability of an event can be expressed as a ratio of two lengths, two time intervals, two areas or two volumes. For example, suppose a painting whose area is 144 square inches is framed by wood 2 inches wide. Someone is painting the ceiling in that room, but doesn't bother to remove the picture. The conditional probability that a drop of paint hitting the picture or frame misses the picture itself depends on W, the width of the picture. Find this probability as a function of W and graph this function. Estimate the dimensions of the picture which has the greatest probability of being hit by a paint spatter.

As a somewhat more sophisticated application, we introduce the idea of finding the area of a region under a curve using Monte Carlo simulations. As one application, we consider the area under the parabola $y = x^2$ from 0 to b. Geometrically, we demonstrate that this region lies inside a triangle using the maximum value of the function, so that its area is less than $\frac{1}{2}b \cdot b^2$. We then perform Monte Carlo simulations to estimate the area corresponding to $b = 1, 2, 3, 4, 5$, say, and use the values so obtained as a set of data to be analyzed. On one set of runs, the data was best fit by the power function

$$A = .333197 \cdot b^{3.0006}$$

with a correlation coefficient of $r = 0.999992$.

In general, our focus is on the *use* of probability as a *tool*, not on a formal treatment nor on standard balls-in-urns problems. We want to demonstrate to students that stochastic ideas and methods provide a valuable tool for investigating nondeterministic phenomena. In turn, we expect that such an applied introduction to the subject will motivate them and will provide the background necessary to make a full probability course more accessible.

Trigonometry is approached from the point of view of modeling periodic phenomena such as the number of hours of daylight in a given location as a function of the day of the year, the tides, the heart and so forth. We put less emphasis on manipulation of trig functions for their own sake. Further, we use the trig functions as a vehicle for exposing students to the ideas of approximating one function by another via "soft" introductions to the notions of Taylor polynomial approximations and Fourier approximations in graphical and numerical settings.

Matrix algebra is introduced not just as a shortcut method for solving systems of linear equations. There is an incredibly rich array of applications of the subject; we show these applications to students in a finite mathematics course, but rarely give even a glimpse of them to students in the mainstream math/science/engineering track. Yet, these are the applications that make the subject come alive to most students. Thus, we introduce matrix algebra as a unifying tool for investigating a wide array of applications including systems of difference equations, Markov processes, and geometric transformations.

Our course also involves much computer or graphing calculator work to explore mathematical models. It involves many classroom experiments to investigate the accuracy of the mathematics in predicting the results of actual processes or to help develop mathematical models based on experimental data. For instance, Newton's Law of Cooling, the damped spring, and Torricelli's Law on fluid leaking out of a cylinder are treated experimentally as well as theoretically.

The course also features a series of student investigations to provide a real-life dimension to the mathematics. For example, we have students conduct individual investigations of mathematical ideas and methods using computer software and/or graphing calculators. We also have students collect sets of data of interest to *them,* say on some growth process or on the acceleration times of a Porsche, and eventually determine the best fitting curve, possibly exponential, logistic or power. Similarly, they have been asked to model the number of hours of daylight in a city of their choice based on actual data collected from the newspapers or other resources.

We find that this direct involvement in the mathematics and the wide applicability of the subject is providing students with the motivation and impetus to continue on to study mathematically related fields. This certainly is reflected in the comments we have received from students in the course. See Appendix II for a set of all the verbatim written comments from the students in one of the pilot versions of the course.

The PreCalculus Reform Project is being funded under a multiyear grant from the National Science Foundation under the joint direction of the author and B. A. Fusaro of Salisbury State University. The initial project working group includes Florence Gordon, Walter Meyer, Jim Sandefur (on leave to work at NSF in 1993), Martha Siegel, and Alan Tucker. We have since added Ellen Shatto and Judy Broadwin to the team to provide expertise on the high school perspective, since we feel that our course would be valuable at that level as well. The project directors have offered pilot versions of the course during the Spring 1992, Fall 1992 and Spring 1993 semesters; broader class testing has begun in Spring, 1993 and will be extended in the coming years. We plan to provide workshops to familiarize instructors with the project materials and prepare them to teach the course.

Appendix III includes a copy of the final exam given by the author during one of the pilot offerings of the course. Despite the sophistication of the problems, the students performed extremely well, particularly in comparison to the author's experiences with more traditional precalculus courses.

Interested readers are encouraged to contact the author or Fusaro for additional information on the details of the project. In particular, we welcome hearing from potential class testers of the course.

Acknowledgement

The work described in this article was supported by the Division of Undergraduate Education of the National Science Foundation grants USE-91-50440 for the Precalculus/Math Modeling project and #USE-89-53923 for the Harvard Calculus Reform project. However, the views expressed are not necessarily those of either the Foundation or the projects.

Appendix I

Table of Contents

(4) Transmission of information

(5) Mechanical systems: simple harmonic motion and damped harmonic motion

(6) Solutions of non-homogeneous difference equations

(7) National income model

(8) Inventory analysis model

(9) Data analysis and higher order difference equations

VIII. Matrix Algebra and its Applications

(1) Matrices and vectors

(2) Scalar products

(3) Matrix multiplication

(4) Gaussian elimination

(5) Matrix growth models

IX. Probability Models

(1) Introduction to probability models

(2) Binomial probability and the binomial formula

(3) Geometric probability

(4) Estimating areas of plane regions

(5) Waiting time at a red light

(6) Waiting for service in a line

(7) The spread of an epidemic

X. Systems of Difference Equations

(1) The predator-prey model

(2) Solutions to the predator-prey model

(3) Richardson's arms race model

(4) Labor-management negotiations model

(5) Lanchester's Square Law for military stalemates

(6) Marriage rules among the Natchez Indians

XI. Geometric Models

(1) Introduction to coordinate systems

(2) Analytic Geometry

(3) Conic sections

(4) The average value of a function

(5) The polar coordinate system

(6) Curves in polar coordinates

Appendix II: Student Survey Responses

"My overall reaction to the course was extremely positive. By emphasizing the value of mathematical pursuits through applications first (and theory being derived from the application), the course proved to be constantly interesting. Past math courses seemed tedious. This course never struck me as tedious (challenging? very, but not tedious) Strong points of the course:

1. The professor's enthusiasm, approachability and good cheer.
2. The small class size made asking questions easy.
3. Most of the examples provided in the text were appropriate.
4. The text was easy to follow (with 1 or 2 exceptions).
5. Above all, being able to see how I could use the subject I was learning in the 'real' world.

Weaker points:

1. The section on Hooke's Law and Newton's Law of springs was tough to follow.
2. Computer material is tough to follow unless the student is the operator. So, my only suggestion would be: every other Thursday have a computer lab. Then, instead of testing us on regression analysis, we could be tested during lab practicals. This grade could count as one question on each exam. Nice to think of math as an intellectual pursuit, a very useful tool, and a way of seeing live—as opposed to 'a course I have to take to complete a chemistry curriculum.' I plan to take calculus through analytic geometry since I'm pursuing chemistry/environmental engineering. My plans have remained the same, but my outlook has certainly improved (By the way, the course somehow managed to cure my phobia of college physics.)"

"I think this course was great. It was probably the most enjoyable math class I ever took. I liked mostly the adding up the numbers in a sequence with $\frac{1}{2}n(n+1)$. It's fun to go home and ask my parents to add up the numbers between one and 1 hundred and they look at me like I'm nuts. And I can do it in a few seconds. Also, the $(1-r^{n+1})/(1-r)$ was very interesting. Also, I liked the Newton's law of cooling stuff like figuring out how long a body's been dead and cool stuff like that. And I am not a kisser but cause my name isn't on this but I have never had any teacher as good as you. And I hope that the next calculus classes are as good as this one. I was always strong at math but this course has made it enjoyable and has given a much more positive attitude towards calculus."

"I liked this course a lot. However, it was a very hard course. We had a very smart teacher that knew how to get down to a student's level and help us understand. It was interesting how much math has to do with real life. Math can really be applied to real life. I see that math is used throughout real life. I wish that my next calculus course would be taught in this way. I am planning to take calculus I and II but now I'm afraid that those courses will be taught in the old way again when I just learned the new precalculus."

"My reactions to this course is a lot better now than it was in the beginning. When I took the previous course I didn't do well. I came into this course with the same attitude, but it has changed. I feel that Difference Equations was the bulk of this course. We did a lot of work for Difference Equations and I now find them fairly easy. Difference Equations are definitely most interesting because of the amounts of time put into it. I also think it has a lot to do with the teacher, I found my teacher very helpful which made the course interesting.

I have always liked math, but the course before this one changed my mind. I didn't like math anymore and found it very hard. But this course changed my mind again. I like math and I find it interesting."

"The sheets were very helpful & followed the structure of the course. Also, I appreciated the fact that the problems were relevant to real life situations. This course was, however, very demanding. I regret that I am taking so many courses this semester and could not allot more time to this one. The work with the computer I found helpful; however, it was confusing to load the programs. ... I can no longer harp that math is irrelevant to the 'real world' because it has been proved to me that it is indeed relevant."

"I felt that this course was very interesting. Prof. Gordon did a great job teaching us the course. Relating math to real life situations is really good. I hate doing word problems, but the way it was represented gave me a chance to do them and maybe somewhat like them. I was always one to just do the problems given the numbers and letters, but this course has helped me be able to figure out an answer by not really having too many numbers and letters. I have always enjoyed math but this course made it just a little bit more fun."

"I wasn't used to the way this class was taught so I found it hard to study for. Answers to the homework problems would definitely be beneficial. I thought the applications to the difference equations were interesting, I liked to see how the mathematics applied to real life. Overall, I liked this class better than any other math class I've taken. This course made me see that math is not just a bunch of numbers and graphs. Math is used in everything even if it is not obvious."

"In a way, I liked this course very much (it was something different), at the same time I believer that I could do better with answers on the back of the book. I think that many things in this course is very interesting and I know that I wouldn't learn them in a regular MA 62 class. I hope more courses will be taught this way and that many students will enjoy it as much as I did."

"I liked the applications to realistic terms, not infinite problems but ones with limits. Finding solutions to difference equations is extremely hard and it went a little to fast. Computer projects are time consuming and frustrating.

In physics we found the data in experiments and gave a quick equation to it. But in here we worked with equations and found experiments that fit. I may be a little ahead now for my science classes."

MA 62 Final Exam

1. a) Find the complete solution to the difference equation

$$x_{n+1} - 3x_n = 12n - 8$$

b) If $x_0 = 2$, find the corresponding solution.

c) Calculate the first 5 terms of the solution using the difference equation.

d) Calculate the first 3 terms of the solution using the solution from part b).

e) Draw a very rough sketch of the solution to show its behavior.

2. Solve each of the following difference equations:

a) $x_{n+2} + 3x_{n+1} - 10x_n = 6 \cdot 3^n$

b) $6x_{n+2} - 5x_{n+1} + x_n = 14$

Draw a very rough sketch of each solution to show its behavior.

3. Given $x_{n+2} - 8x_{n+1} + 20x_n = 0$.

a) Find the general solution.

b) What are the modulus and angles for the complex roots?

c) Convert the general solution into the equivalent solution in trigonometric form.

4. A certain radioactive element decays 20% each year. If 25 grams are present initially,

a) what is the amount present after 10 years?

b) what is the half life?

5. The kidneys remove 80% of any impurity from the blood each day. A woman is prescribed a daily medication with a dose of 100 mg each day.

a) Solve the difference equation for the level of drug in her system daily.

b) What is the level of dosage in her system after 10 days?

c) What is the maintenance level of the dosage?

6. The United States produced 2.5 million barrels of crude oil in 1985 and production has been increasing at an annual rate of 1.7%. If this trend continues, what is the total U.S. oil production between 1985 and 2000?

7. A blacksmith takes a horseshoe out of a furnace at 1200° and dunks it into a barrel of water at 70°. After 10 minutes, the temperature of the horseshoe is 200°.

a) What is the temperature after 15 minutes?

b) How long does it take until the temperature reaches 80° so that the blacksmith can take it out of the water and put the shoe on a horse?

(According to Newton's Law of Cooling: $T_n = (T_0 - R)(1 + \alpha)^n + R$)

8. A young couple sets up a savings plan based on a 4% annual interest rate. They deposit \$1000 initially and agree to increase the annual contribution by 10% each year.

a) Express the above situation as a difference equation.

b) Find the solution to this difference equation.

c) Sketch a very rough graph of the solution to show its behavior.

9. Given the points $A(1, -2)$ and $B(7, 16)$.

a) Find the equation of the line through A and B.

b) Find the midpoint of the line segment from A to B.

c) Find the point which is 3/5 of the way from A to B.

d) Find the distance from A to B.

10. The following graph shows the velocity of a plane as a function of time.

Indicate the following:

a) all intervals where the velocity is decreasing.

b) all intervals where the plane is accelerating.

c) all intervals where the curve is concave up.

d) all intervals where the curve is concave down.

e) all points where the velocity is greatest.

f) all points where the velocity is increasing most rapidly.

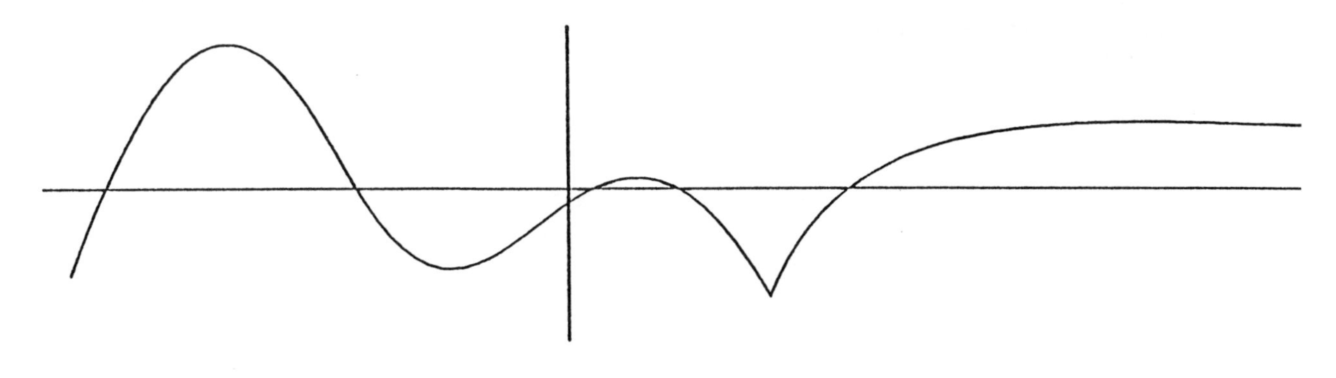

11. At a certain pier, the low water line is 8 feet above sea bottom and the high water line is 14 feet above bottom. If low tide occurs at midnight and high tide at 6 A.M., then the height of the water as a function of time t is $H = 11 + 3 \cdot \sin(2\pi(t-3)/12)$.

a) How high is the tide at 10 am?

b) When is the water 12 feet deep?

12. The perihelion and aphelion distances for Mercury are 35.4 and 36.0 million miles, respectively. Find the equation of Mercury's orbit about the sun.

13. Given $z = \|z\|(\cos\theta + i\sin\theta)$ and $z^2 = \|z\|^2(\cos 2\theta + i\sin 2\theta)$, show that $z^3 = \|z\|^3(\cos 3\theta + i\sin 3\theta)$. (Hint: $z^3 = z \cdot z^2$.)

14. A biologist is attempting to develop a mathematical model relating a person's height (in inches) to his or her weight (in pounds) from infancy to maturity. She suspects the relationship involves a power law. The data values for one person studied are:

Wt:	5	17	24	33	46	59	73	98	120	195
Ht:	19	26	31	40	46	52	57	66	70	75

a) Draw the scatterplot for this data and indicate the regression line by eye.

b) Transform this data to linearize it.

c) Draw the scatterplot for the transformed data and indicate its regression line. Suppose the computer produces the following linear regression equation for the transformed data:

$$Y = .43X + .95$$

d) De-transform this equation to produce the exponential function which best fits the data.

e) What is your prediction for the person's height when he weighed 140 pounds?

f) What is your prediction for the person's height if he weighed 220 pounds?

g) Which prediction would you have more confidence in? Why?

15. Transform $\sin^4 x$ into an equivalent expression that does not contain any powers.

"Are You Ready?" Disks: an Aid to Precalculus Reform

David O. Lomen and David Lovelock
University of Arizona

Introduction

An alarming number of students start their first collegiate mathematics course without adequate mastery of prerequisite material, even though they have covered this material in previous courses. Two of the most common responses of collegiate instructors to this predicament are:

- to begin the course with a few (or several) lectures covering prerequisite material; or
- to begin the course with new material, assuming their students will eventually catch up.

In this paper we propose an alternative approach, designed to avoid the obvious drawbacks of these two responses, namely,

- to provide students the opportunity to review appropriate prerequisite material, at their own pace, as it is needed in the course. (This opportunity is provided for students to use outside of class meetings.)

The University of Arizona has developed computer programs to address the problem of inadequate mastery of prerequisite material for several freshman and sophomore mathematics courses.

Description of the "Are you ready?" disks

In 1987 the Mathematics Department at the University of Arizona developed the "ARE YOU READY FOR CALCULUS?" computer program. This program reviews only those aspects of prerequisite material needed to succeed in traditional first semester calculus. It contains multiple choice questions which are divided into ten different areas from algebra and trigonometry. Available at all times are HELP screens, which are different for each area. Each HELP screen contains a list of formulas needed to solve the problems posed in that area. Thus the student has access to the needed information, but it is not pointed out to the student *which* formula is required. Besides serving as a check on the student's retention of the material, the HELP screens provide a convenient reference for students to access important material. It is certainly easier to find this material on a disk than to locate the appropriate book (and the proper page) used from anywhere from three months to three or more years ago.

After ten questions have been attempted, the program provides a diagnosis of the student's performance, and if substandard, lists a section in an appropriate Schaum's Outline for the student to study and specific exercises to work before attempting more questions on the computer disk. These suggested sections and exercises contain material where ideas and examples are found appropriate to the weakness revealed by the diagnosis.

Student response to this initial disk was very positive, so we continued to develop "ARE YOU READY FOR __ ?" disks for other mathematics courses. To date we have disks for Intermediate Algebra, College Algebra, Calculus I, Calculus II, Calculus III, and Differential Equations. We are in the process of developing ones for Statics, Advanced Engineering Mathematics, and Advanced Placement Exam in Calculus (AB and BC).

The instructions for the "ARE YOU READY FOR __ ?" disks offer the following advice for students to get the most out of its use.

1. *Always* use pencil and paper.
2. *Always* start from the question and work towards an answer. *Do not* start from the answers and work towards the question—some questions can be done this way in this review, but that won't help you solve new problems.
3. Remember this review is for your benefit – take it seriously. If you don't understand something fully, admit it, *and do something about it!*
4. If you don't know why you missed a question, copy it down, along with all the possible answers, and take it to your instructor for help. Most instructors are

human and are quite happy to help a serious student.

Ways of using the "ARE YOU READY FOR __ ?" disks

The first day of class we announce that there will be an hour exam one week hence covering prerequisite material. All the questions on the exam will be over the material contained in the appropriate "ARE YOU READY FOR __ ?" disk. (We make this exam count towards their final grade in the course to insure that students take this exam seriously.) This strategy has several advantages:

- It gets the students working immediately and reminds them of the importance of prerequisite material.
- It makes the students aware of any weakness they have in prerequisite material and provides a means for them to correct these weaknesses.
- It allows totally unprepared students to change to a prerequisite course in time to succeed in that course.
- Since most classes do not have exams the second week of the term, all subsequent exams are usually out of sequence with exams in the students' other courses.
- It permits people who have been away from the material for some time an efficient means of getting up to speed.
- It provides a way for students with a prior "watered down" course to rectify the situation. (For example, a student from a very small school said "I never knew whether or not my high school was any good. Now, with this disk, it doesn't matter.")
- It saves instructor's time during class because it is no longer necessary to teach prerequisite material as part of the course.

Many students, especially those who have returned to school after several years absence, find it extremely helpful to use the disk while doing homework. The disk provides quick access to HELP when the student knows roughly what to do but has forgotten the details. Having HELP available as they do homework contributes to effective learning much more than postponing the working of a troublesome exercise until the student's professor can be seen during office hours.

Some of our mature students credit these disks with their success in mathematics courses that formerly were very difficult for them. It also reinforces the idea that the students' performance in a class is mostly dependent on their effort, and they are responsible for overcoming any previous shortcomings in their background.

Students have ready access to these disks since they may copy the "ARE YOU READY FOR __ ?" disks and use them on their own, or any other computer. The programs are in the public domain and may be freely copied and distributed.

The first version of "ARE YOU READY FOR CALCULUS I?" was released to the 1987 Summer Session students, and one of the sections was surveyed at the end of the course. Ninety-four percent of these students used the disks and reported the following usage:

Less than 1 hr:	0%
Between 1 and 2 hrs:	25%
Between 2 and 3 hrs:	12%
Between 3 and 4 hrs:	25%
Between 4 and 5 hrs:	13%
More than 5 hrs:	25%

To the question "How do you feel about its usefulness/helpfulness?," the students answered:

Very useful:	63%
Useful/helpful:	37%
Ho-hum:	0%
Not useful/helpful:	0%
Waste of time:	0%

To the question "What do you think of the idea behind this project?", the students responded:

"Needed."

"Absolutely a great idea."

"An excellent idea. It really brings out weak areas."

"Good idea. Lets a student know where he or she stands."

"I feel that this project is an extremely good idea."

"Great idea."

"Groovy."

"Brilliant."

"It's a wonderful idea; saves time, money, hassle."

In response to "Did you enjoy using the program?", the students wrote:

"No. It pointed out a few too many weak areas—which is good."

"I enjoyed the program."

"Yes. The program will come in handy for the next 4 years."

"Yes. One can jump right into it."

"I was very glad to be made aware of the areas in which my preparation for calculus was lacking."

"Yes. It helped me to feel I could get started in the right direction."

"Yes, very much. It helped me in areas I had forgotten."

To the question "Please list anything you particularly liked about the program.", the students answered:

"I liked the choice/comment/command format. I also like your comments that come on after a student answers a question. I like the intro screen with the clicks, the quotes and the presentation of the problems."

"I liked everything."

"Sense of humor."

"The hints after the comment."

"Simple."

"The length of each quiz."

"It helps a student to be more interested in learning Math with the use of a computer. The quizzes are short but it points out the important aspects in each topic."

In response to "Any suggestions/comments?", they wrote:

"The program is good for the students."

"More programs like this."

"I took Business Calculus from Dr. Lovelock in 1982. I failed the first quiz and he suggested I drop. I struggled through and got a C but it was unfortunate this wasn't available then."

"Overall, this is a super program and a great idea. I think this program would be especially good for students coming back to math after a year, semester, or summer with no math classes. It would serve as an excellent review. It is easy to use and to read, and the problems are not really that difficult. I think this program should definitely be distributed to other colleges, etc."

We have obtained similar enthusiastic responses from students in the other courses for which "ARE YOU READY FOR __ ?" disks are available. The University of Arizona is not the only school involved with this endeavor. Over 700 academic institutions in Argentina, Australia, Austria, Belgium, Canada, Columbia, Cyprus, Denmark, England, France, Japan, Malta, Mexico, New Zealand, Nicaragua, Norway, Pakistan, Poland, Scotland, Sweden, Switzerland, the United States, and Wales have obtained copies of our programs. For example Arizona State University uses the "ARE YOU READY FOR CALCULUS I?" disk to place students in the proper class. Embry Riddle Aeronautical University in Prescott sends a copy of this disk to every incoming freshman to "bone up" on precalculus ideas so they can successfully start their calculus course fall semester, since they have no precalculus course there.

The six disks,

ARE YOU READY FOR INTERMEDIATE ALGEBRA?
ARE YOU READY FOR COLLEGE ALGEBRA?
ARE YOU READY FOR CALCULUS I/ BUSINESS CALCULUS?
ARE YOU READY FOR CALCULUS II?
ARE YOU READY FOR CALCULUS III?
ARE YOU READY FOR DIFFERENTIAL EQUATIONS?,

along with thirty-seven other public domain software packages we have developed for the MS-DOS environment, may be obtained by downloading on Internet by ftp (file transfer protocol). The list of available software and instructions for downloading are found in the file README on math.arizona.edu. Information on other means of obtaining the software, as well as a software manual, is found by writing "Mathematical Software, Mathematics Department, University of Arizona, Tucson, Arizona 85721." The last five disks in the "ARE YOU READY FOR __ ?" series listed above are available for the MacIntosh (for both System 6 and System 7).

Connection with precalculus reform

"ARE YOU READY FOR __ ?" disks can diagnose student weaknesses in material prerequisite to any precalculus course. (If the questions we have on a disk are not appropriate for your situation, we will be happy to work with you to develop ones that are.) These disks allow the student to practice previously learned skills without having the instructor use class time to review. For example, the "ARE YOU READY FOR CALCULUS I?" disk includes the following sections: Elementary Trigonometry, Elementary Algebra, Factoring, Fractions, Functions & Equations, Inequalities, Curves, Functions from Formulas, and Trigonometric Identities. The disks also provide information in a convenient format for the student to use as an aid while doing their homework. The instructor is able to present a new approach to precalculus without having to take time to review ideas students should have mastered from previous courses. Time saved from this could also be used for more cooperative or exploratory learning situations. (In fact we encourage students to work together on these disks.) Since this software is in the public domain, it may be made available to the student for the cost of the disk. This also means the students can use the disk at any machine available, either on or off campus. Thus, mathematics departments need not have extensive computer facilities (or any at all) in order for their students to benefit from the use of these disks.

Pacesetter Mathematics

Carolyn R. Mahoney
California State University San Marcos

Introduction

In the Fall of 1991, the College Board launched Pacesetter Mathematics [1] as an educational reform effort aimed at encouraging more students to take four years of high school mathematics by providing a rich and rewarding capstone experience. As recommended by NCTM's *Professional Standards for Teaching Mathematics* [4], Pacesetter seeks to close the gap between recommended ideals of teaching mathematics and the reality of mathematics education in the schools of today by providing secondary school course frameworks and related assessments, as well as professional development opportunities and other services and products that will help embed Pacesetter mathematics in the practice of schools and teachers. The Pacesetter course framework includes a topic structure and content emphasis as well as a set of learning and assessment activities that serve to promote and improve learning. In this context, the Pacesetter assessment process seeks not simply to measure student achievement, but also to inform instruction and invigorate learning. To promote change, Pacesetter will merge professional development activities with the everyday work of teachers in the classrooms by including detailed teacher notes with the framework, by assisting teachers with annotating current textbooks and course materials to integrate Pacesetter, by establishing local opportunities for networking and exchanging ideas, and by providing inservice workshops in the school or district. Pacesetter is, therefore, the first school-level College Board program to move ambitiously beyond prediction in support of comprehensive educational improvement.

The Pacesetter Task Force that is overseeing the development of the instructional support materials and the classroom and end-of-year course assessments, and the design and implementation of the professional development activities is co-chaired by John Dossey (Illinois State) and Phil Curtis (UCLA) and includes university mathematicians and mathematics educators, community college instructors, and high school teachers; the author is a member of this broadly representative committee. The course framework has been reviewed by members and staff of the National Council of Teachers of Mathematics and the Mathematical Association of America, each of which is represented on the Task Force. Industry representatives serve as consultants to the Task Force and regularly attend and contribute at Task Force working meetings. The course was piloted diring the 1993–94 school year in seven sites that represent a broad cross section of ethnic, racial, and geographic diversity—Broward County, Florida; Prince George's County, Maryland; Irving, Texas; Battle Creek, Michigan; San Diego, California; Charlotte-Mecklenburg, North Carolina; and Rutland, Vermont [2].

The Course

The Task Force had several goals in mind as it discussed course content. The over-arching goal is to provide a capstone learning experience for high school students who have taken at least three years of high school mathematics including Algebra II. Students who successfully complete the course should be well prepared for college level mathematics including the calculus. These students should have had enough experience working with data and developing mathematical models to go directly into quantitatively based college courses not requiring calculus, in particular beginning science courses, probability and statistics, and finite and discrete mathematics. Through the analytical, statistical, and modeling work contained in Pacesetter, course completers should be prepared to handle technical problems that they will meet if they go directly into the work force. The course strives to develop on the part of the student an attitude of independence toward mathematics and a willingness to confront new and unexpected problems by requiring students to work in groups to analyze problems, carry out investigations, and use grounded mathematical models. Special attention is given to the development of oral and written mathematical communication skills in the individual student activities and group projects. It is intended that students and teachers become active participants in the work of the course and in the development of major themes in the syllabus [3].

The Pacesetter framework is presented as a flexible plan of study, not a fully prescribed curriculum. The themes of this course are mathematical modeling involving optimization, prediction and data analysis, the use of technology as a tool for deeper understanding of mathematics, and the development of mathematical structure and symbol sense. The course itself is organized into six major units of study that develop understanding of the nature of functions, their properties, and their applications [3]. The six units are: linear functions, exponential functions, polynomial and rational functions, trigonometric functions, modeling with matrices, and modeling with more general functions. The emphasis throughout is that the mathematical ideas should be introduced in a problem context. These applications may be motivated by consideration of real life quantitative situations involving data or from questions arising in science and mathematics. Students are required to confront, investigate, and resolve problem situations containing visual, numerical, symbolic, and semantic information. Students gain experience in selecting, implementing and defending appropriate solution strategies, and in communicating the solutions and related decisions in the context of the situations studied.

For example, one of the Task Sets in Unit 6 may be built around linear recurrence relations. It is assumed that all students have access to a TI-81 or TI-82 or similar calculator. The class will be divided into three parts, and each part works, in groups, on one of the following three situations:

1. A certain fish farm starts out with a population of 6,000 fish. The fish population grows at a rate of 5% a year (births outpace deaths, no poaching takes place, etc.) Also, a batch of 1,000 fish is added to the fish farm every year.

2. Suppose that at noon you take an initial dose of 30 mg. of some medicine, and then take a dose of 15 mg. at noon every day after that. Suppose further that your body removes 25% of the medicine that was in it at the start of the 24 hour period.

3. You take out a loan of $3,000 on December 1 and agree to repay it by making a payment of $75 on January 1 and every month thereafter until the load is paid off. The interest rate is 18% per year, compounded monthly.

In each of the situations students are asked to make tables, draw graphs, and give written descriptions, and discuss the long-term behavior of the system. Through written questions and classroom activity students develop a linear recurrence relation for their situation. Students present their group finding and through whole class discussion notice the similarity in the findings and move to develop the formula $a_n = Ra_{n-1} + B$. They return to small groups to think of a situation which might be described by this general recurrence relation when $R < 1$ and $B < 0$. Curious students realize that there are seven cases possible. They are asked to discuss the significance of the initial condition in each case, are encouraged to explore what happens when the initial condition is changed, and are asked to generate the first eight members of the sequence arising from having chosen some specific recurrence relation. Since this is Unit 6 and students have studied arithmetic and geometric sequences earlier in the course, students are given activities which lead to the development of the closed form formula

$$a_n = AR^n + B\frac{1-R^n}{1-R} = (A - \frac{B}{1-R})R^n + \frac{B}{1-R}.$$

Based on the closed form expressions for the seven cases, students are asked to construct and discuss graphs for each case. Through experimentation and discussion, they notice, for example, that in some cases the graphs increase and then level off to the equilibrium value $\frac{B}{1-R}$. They discuss the significance of the graphs' behaviors in the context of the original problems; for example, they discuss the case when medicine is added in discrete amounts when the pill is taken, but is continuously removed by the body. Finally, they do a "cobweb analysis" on these graphs. Students then are given other situations for which they develop and discuss the attendant recurrence relation.

Conclusion

In summary, Pacesetter is a comprehensive effort for increasing the number of students who elect to take four years of high school mathematics. Pacesetter seeks to bring about change by providing a rich curricular framework and assessment package. Further, the Pacesetter effort seeks to provide the support for teachers that is crucial to its full and successful implementation. Specifically, the Pacesetter package will include: the course framework described above including copious teacher notes containing background motivation and theory; strong teacher inservice and support activities including the course framework, in-school assessment exercises, summer workshops and institutes, and publications illustrating successful teaching activities; classroom assessment that permit teachers to monitor and shape instruction, while providing feedback to students; end-of-course achievement tests, including both multiple choice and performance-based sections; and a rigorous system for scoring Pacesetter's end-of-course achievement tests on a state, regional, or local level, including inservice for teachers who score them [1]. This substantive attention to the challenges of implementation should help schools, school districts, and states provide the continuous preparation required to unite educational equity with excellence.

References

1. The College Board, "Pacesetter," College Board, 45 Columbus Avenue, New York, October, 1991.
2. The College Board, "The Pacesetter Letter," College Board, New York. Winter 1993.
3. Dossey, John, "Pacesetter Mathematics, College Board Private Report. New York, 1993.
4. National Council of Teachers of Mathematics, *Professional Standards for the Teaching of Mathematics,* Reston, VA, 1991.

Improving Success in Calculus

Jerold Mathews and William Rudolph
Iowa State University

Keith Stroyan
University of Iowa

Introduction

Our aim in the Improving Success in Calculus (ISIC) Project is to increase student success in calculus. Toward this end the course that we are developing to precede calculus focuses on applications-based problems that shape the course and develop the necessary problem solving skills. This course follows the recommendations of the NCTM and MAA regarding mathematics courses for science students in grades 11 or 12 as preparation for science and calculus. Specifically,

- the course is focused on its goals,
- problem solving is a major theme,
- problems from industry and science are used, and
- appropriate use is made of available computing technology.

The ISIC Project prepares students for entry into the twenty-first century workplace. The mathematics of our proposed course not only prepares students for calculus, it is useful as well, as demonstrated by our real applications. Moreover, the methods of presentation develop life-long skills. Specifically, we stress

- ability to set up problems,
- knowledge of various techniques to solve problems,
- ability to work with others to solve problems, and
- ability to break complex problem into smaller ones.

A major goal of our course is to prepare students for calculus. To this end ISIC is

- working with high school teachers to restructure the course preceding calculus, to improve preparation for calculus (as measured by university mathematics placement and longitudinal studies on success in university calculus),
- working with teachers and students to make problem solving a major theme,
- collecting applications-based problems with the assistance of business, industry, and science professionals, and refining these problems for classroom use,
- focusing on non-lecture oriented student learning, including cooperative learning,
- integrating computing technology into the course, primarily graphics calculators, as a natual adjunct to problem solving, and
- developing materials for a book to be titled something like *Vector Geometry and Trigonometry in the Real World*, VG&T for short.

We believe more students will succeed in calculus if they learn the prerequisites in an environment structured around solving problems arising in mathematics or its applications. The problems should motivate the prerequisite topics and justify the expectations we have of students. We believe that such an environment helps students to value mathematics, reason mathematically, communicate mathematically, develop confidence in their abilities, and become mathematical problem solvers. These are the five general goals for all students given in the NCTM *Standards.* ISIC recognizes that the best place for science students to learn precalculus mathematics is in high school, in a non-remedial environment, with highly trained teachers and adequate time.

Collateral Efforts

Prior to focusing our efforts on the ISIC Project, we and many others nationally have worked to improve the success at the university of college-intending students and, more broadly to prepare all students mathematically so they may become productive citizens in an increasingly technological society. Iowa is participating in the current efforts to implement the NCTM *Standards.* A committee consisting of mathematics education leaders in the state is conducting workshops throughout Iowa to make teachers aware of the *Standards* and to assist in their implementation in the schools. Iowa has submitted to NSF a plan for systemic change in mathematics and science education. Finally, over the past decade the state universities in Iowa have used mathematics placement tests during orientation to improve the success of freshmen in their first mathematics course. It appears, however, that this practice had little effect upon high school preparation. Its main effect has been to increase the enrollments in non-credit mathematics courses at the university level.

Three years ago the three Iowa state universities, several private colleges, and the state community colleges jointly decided to try a pilot early mathematics placement testing (EMPT) program, based upon the Ohio model. Since this testing is usually done during the junior year, EMPT has more promise for improving preparation for university mathematics than does testing during summer orientation. After a decade of testing, Ohio State University has cut its remediation rate in half.

While we are waiting for our own 50% reduction in remedial enrollments and for the effects of the NCTM *Standards* on the preparation of students for college mathematics, students and faculty members continue to experience the demoralizing effects of requiring up to two-thirds of the students in some majors to delay calculus for a semester or more to improve their mathematics background. The ISIC Project aims to improve the current situation, which in our opinion is independent of whether the calculus is traditional or reform.

ISIC

Two years ago ISIC was organized at Iowa State University (ISU) under a grant from the Department of Energy (DOE). In November 1992 the DOE eliminated funding for all mathematics projects in the United States, and we sought and found interim funding for a reduced 1992–1993 program and are presently seeking funding through NSF. ISIC is a joint project with mathematics faculty members from the three state universities in Iowa, DOE scientists, university and industrial engineers, and high school teachers. During the first year a series of panels, talks, and discussions with a group of 15 high school teachers from the Des Moines area were held. We explored existing pre-calculus courses in the six high schools represented, planned how the group might go about improving success in calculus by changes in the course preceding calculus, and heard talks from engineers (from Maytag, John Deere, the American Institute of Architects, and Iowa State University College of Engineering), scientists (from the Ames Laboratory of the DOE), and mathematicians (from Iowa State University). The talks focused on applied problems using VG&T methods and stressed the importance of good preparation in these topics for university work.

From these activities we began to work out a plan designed to improve success in calculus in Iowa and, more broadly, in schools across the United States. In the second year colleagues from the University of Iowa (UI) and the University of Northern Iowa (UNI) joined the ISIC Project and 10–12 leading high school teachers were selected from across Iowa. Jointly we are

- developing and refining applications–based mathematics with project teachers,
- assisting teachers in classroom testing of these materials in schools, and
- providing support to project teachers and schools through visits to classrooms, discussions with teachers, workshops, development of materials, and evaluation activities.

The ISIC Project is about to enter its third year. We have collected some 40 problems having potential to become key problems for our planned two-semester course. Samples of these problems are given in the Appendix. Most of the problems assume students have easy access to a graphing calculator. Other problems are being sought to strengthen the applications-rich environment, problems which can motivate both the topics and problem-solving skills necessary for success in calculus and help set the level of expectations of the course. We have also developed a first draft high school level treatment of 3-dimensional analytical geometry. This grew out of several summer teacher workshops on geometry and computer graphics at UI. Many problems of the real world involve more than two dimensions. Vector methods can be used to develop both two- and three-dimensional analytical geometry with little additional effort in two dimensions, but a major pay-off in three, which in turn opens the way to a host of possible applications.

Each of the 10–12 teachers is preparing a "unit" of material. A unit is a coherent 2–3 week segment from trigonometry or analytic vector geometry (VG&T). The unit should begin with one or more solid applications from VG&T to a physical, biological, engineering, or mathematical science problem. Ideally it should guide the students to the topic or topics of the unit and its solution should serve as a goal of the unit. Progress is being made towards this goal. We are working with the teachers individually, visiting their classrooms twice during Spring, 1993 and with them are revising the unit they have chosen. Student teams are working enthusiastically on problems and in presenting their solutions. At this point we are encouraged by student response to demanding, applications-based problems.

Directions for 1993–1996

The long-term goal for the ISIC Project is to develop materials for a precalculus course including trigonometry, 2- and 3-dimensional analytic geometry with vectors, and parametric representation of curves. The graphing and vector topics will be integrated into the course, not relegated to a final chapter. The course will have a strong orientation towards student solution of non-routine, applications-based problems, and will make appropriate use of computer technology. It will be implemented in 12 lighthouse schools across Iowa and provide direction for a national effort toward implementation of an applications-based precalculus course. It is anticipated that most students will, upon completion of this course, enroll in a mainline university-level calculus course. Students from our course will prime the calculus pump and increase the number of students choosing and remaining in the engineering, physical and mathematical sciences. To accomplish this goal we will continue major development and testing of curriculum materials for the ISIC Precalculus course. Curriculum materials will include those previously developed at the University of Iowa in their NSF-funded vector geometry project, materials previously and currently being developed in the ISIC DOE funded project, commercial materials like those developed at the Carnegie Foundation North Carolina School Science and Mathematics Project, and other applications-based materials like those appearing in the NCTM applications yearbooks and National Aeronautics and Space Administration publications.

We believe the continuing cooperative efforts of the three Regents' institutions in Iowa can have a major impact upon the precalculus curriculum in Iowa. Success in Iowa will produce materials widely useful throughout the United States. Several activities will be the focus of our efforts during this time period. We anticipate continuing cooperative efforts by participating teachers, school administrators, business and industry leaders, scientists and university faculty members. Professionally made video tapes currently under development will be available. These tapes will describe the importance of mathematics and science and will publicize the ISIC Project and its curriculum. The tapes will be shown to parent groups, administrators, teachers, and other interested in education. We plan to start the tapes with a brief discussion by a well known scientist, have a segment on an overview of the project, lead into actual presentations of project materials in classrooms and conclude with a panel discussion on the importance of mathematics and science in careers in engineering, mathematical and physical sciences.

Our efforts towards determining university expectations of what students need in their precalculus course will be continued and refined. Our initial list of essential topics for precalculus came from workshop participants and curriculum committees at ISU, UNI, and UI. A copy of their recommendations is available from any of the authors.

Teachers from the lighthouse schools will attend academic year and summer workshops in which issues in the precalculus curriculum will be discussed and further problem development and refinement of materials will occur. Implementation and refinement of these materials in project classrooms will continue. Support of project teachers will be provided by university professors visiting the project classrooms to observe and discuss concerns. To summarize, our current plans, subject to obtaining long-term funding, include assembling a one semester collection of units for classroom trials in 1993–1994, revision and extension to a two semester collection of units for classroom trials in 1994–1995, and national testing of a preliminary edition of the planned VG&T textbook in 1995–1996.

Problems and Problem Solving

The NCTM *Standards* has as a goal that students should become problem solvers. What factors make this goal attainable? Leaving aside such important global factors as consistent parental support and guidance, self-esteem and confidence on the part of the student, a competent, confident mathematics instructor, and a school administration which supports teachers whose goals are consistent with the NCTM *Standards,* we list several important local factors.

1. Problems rooted in mathematics or its applications should drive the syllabus and set the tone of the course. If such problems do not exist, it is likely that the syllabus is inappropriate.

2. The teacher and text should consistently arrange for, encourage, and expect student "ownership" of problems. For this we must replace the large numbers of elementary, template problems with which students now so frequently fill their time. Problem assignments should always include problems not yielding to follow-the-example behavior, but which require some genuine engagement with the problem or text.

3. The teacher should regard problems and the process of solving them as a major component of the course. He or she should project confidence and enthusiasm for problem solving. As far as possible, problem solving should drive the course and set the standards.

4. Problem selection, presentation, and solution should use the computing technology available to the student.

5. Problem solving provides a strong context for learning to communicate a problem and a solution to others. Written and oral communication to others of ones work is an important and necessary part of learning mathematics.

APPENDIX

Retrograde Motion of Mars

Introduction If at each midnight for the next several years you were to observe the position of Mars against the background of the fixed stars, you would find that its angular position along the ecliptic increases a little each night, but that from time to time it decreases for several weeks. As seen from Earth, the motion of Mars is said to be either direct or retrograde, depending on whether its angular position is increasing or decreasing.

The retrograde motion of Mars was observed by ancient astronomers. The following careful description of the retrograde motion of Mars was taken from a Babylonian clay tablet found in the ruins at Nineveh. "Mars at its greatest power becomes very bright and remains so for several weeks; then its motion becomes retrograde for several weeks, after which it resumes its prescribed course."

In this problem we ask you answer several questions concerning retrograde motion. We make several simplifying assumptions about the motions of Earth and Mars. These assumptions do not change in any essential way the phenomenon of retrograde motion. We assume

- The plane of Mars's orbit coincides with the plane of the Earth's orbit. (The actual plane is tilted 1.8° from that of Earth.) The intersection of the plane of the Earth's orbit with the *celestial sphere* is called the *ecliptic*. The celestial sphere may be regarded as an observer-centered sphere. It is the "background of the fixed stars." Its radius is very, very large relative to Solar System distances.

- The orbits of Mars and Earth are Sun-centered circles. (The actual orbits are ellipses, with eccentricities 0.093 and 0.017, respectively. The eccentricity of a circle is 0.) We use the Astronomical Unit (A.U.) as our unit distance. A distance of 1 A.U. is defined as the mean distance from Earth to Sun ($\approx 1.496 \times 10^8$ km). The radius of the Earth's orbit is 1.0 A.U. and that of Mars 1.52 A.U..

- We assume there was a time, which we take as $t = 0$, at which Mars was on the line joining the Earth and Sun and, in particular, on the same side of the Sun as Earth.

- The period of Mars is 1.88 Earth years (Y). We take 1 Y as our unit of time.

From these assumptions and general background knowledge, we may write vector/parametric equations of the orbits of Earth and Mars. In problem 1 below we ask you to assign proper values to the constants in the following equations:

$$\mathbf{r}_E = \mathbf{r}_M(t) = r_E\,(\cos\omega_E t, \sin\omega_E t)\,, \quad t \geq 0$$

$$\mathbf{r}_M = \mathbf{r}_E(t) = r_M\,(\cos\omega_M t, \sin\omega_M t)\,, \; t \geq 0$$

1. Assign values to the constants in the vector/parametric equations above. Graph the two orbits on the same set of axes.

2. For any fixed $t \geq 0$, the line with vector/parametric equation

$$\mathbf{r} = \mathbf{r}(u) = \mathbf{r}_E(t) + u\,(\mathbf{r}_M(t) - \mathbf{r}_E(t))\,,\; u \geq 0$$

pierces the celestial sphere at a point S_t, while the line with vector/parametric equation

$$\mathbf{r} = \mathbf{r}(v) = v\,(\mathbf{r}_M(t) - \mathbf{r}_E(t))\,,\; v \geq 0$$

pierces the celestial sphere at a point S'_t. Explain why the angular position of Mars as seen from Earth may just as well be measured using the point S'_t. You may wish to use a diagram as part of your explanation.

3. If **a** and **b** are vectors, the angle $\theta \in [0, \pi]$ between them may be calculated from

$$\theta = \arccos\left(\frac{\mathbf{a}\cdot\mathbf{b}}{\|\mathbf{a}\|\,\|\mathbf{b}\|}\right)$$

We have plotted in Figure 1 the angle θ between the vectors $\mathbf{r}_M(t) - \mathbf{r}_E(t)$ and $(1, 0)$, $0 \leq t \leq 3$. Explain this graph.

4. Reproduce Figure 1.

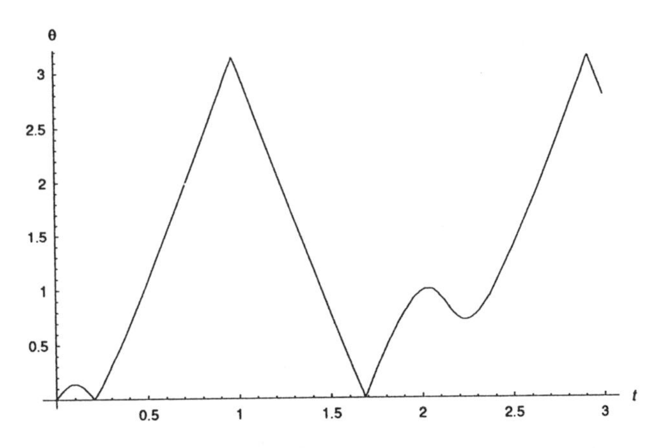

Figure 1: Angular Position of Mars on the Ecliptic as Seen from Earth.

5. Find the times $t \in [0, 3]$ at which Mars is brightest and dimmest. Explain.

6. Explain the comment taken from the Babylonian tablet.

Beam Problem

What is the longest beam that can be moved around the corner shown in the figure? The two corridors are 8 feet and 5 feet wide and the beam is 1 foot square. Due to the weight of the beam it must be kept horizontal during the move. We show in the figure a beam at a typical point during its move.

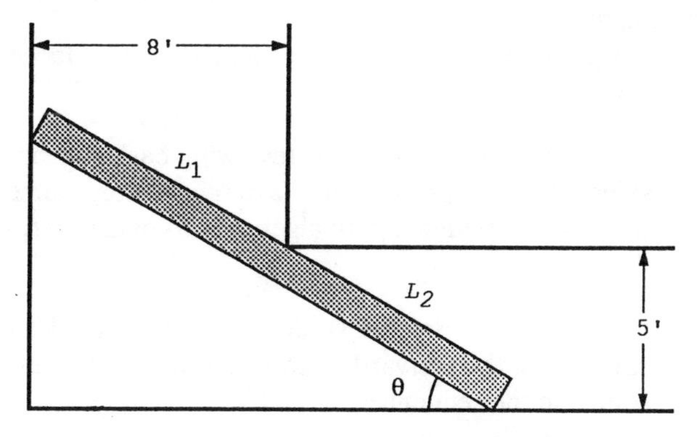

Figure 2: Beam rounding the corner.

One way of approaching this problem is to express the length L of the beam as the sum $L_1 + L_2$ of the two lengths shown in the figure and then to express each of them in terms of θ. To find the longest beam, find the smallest value of L as θ varies. Your graph of L against θ should resemble the graph in Figure 3.

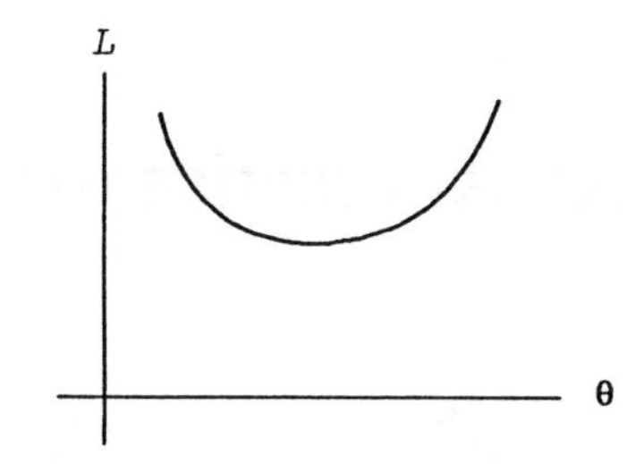

Figure 3: L versus θ.

Board Room Table Problem

Introduction A carpenter is building a table for a corporate board room, following the plans shown in Figure 4. The center of the table is inlaid marble and the border is walnut. The top curved edge of the table is an arc of a circle C. The edge L of one of the pieces of marble is a segment of a radius of C and makes an angle of 7.5° with the end of the table.

Find the center width of the table and the radius of C. From these the carpenter can calculate the width of the walnut boards required for the long edges of the table and scribe them for the required cuts.

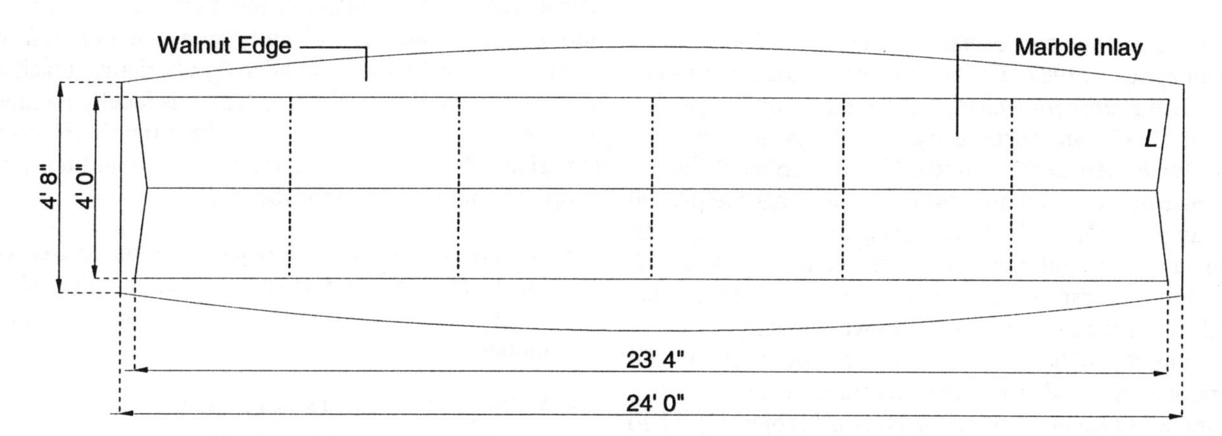

Figure 4: Corporate Board Room Table

An Alternative to the Traditional Precalculus Course

Elias Toubassi
University of Arizona

Background

Students at the University of Arizona who want to take beginning mathematics courses must take one of two readiness tests. Currently we use tests developed in California by the Mathematics Diagnostic Testing Project. The tests primarily cover topics from algebra, geometry, and trigonometry. Some of the students who end up in our precalculus course are ones who had hoped to start in first semester calculus. Most, if not all, have had four years of high school mathematics ending with a course equivalent to precalculus, and in some cases they have had calculus. Thus these are students who had followed the advice of their teachers and counselors by meeting or exceeding the admission requirements of the University. These students represent the second tier as measured by our readiness test.[1]

Precalculus Course

Our precalculus course has been fairly traditional since its inception in 1987. The book we have used for the most part is *A Primer for Calculus*, 5th edition, by Holder [2]. (In 1991–92 some sections used *Contemporary Precalculus Through Applications* by the North Carolina School of Science and Mathematics [4].) Two years ago we piloted a loan program of TI-81 graphing calculators to students. This proved popular with students and faculty alike. After three semesters the loan program was discontinued, and students now purchase their own calculators. I came into this course having been influenced by several recent experiences. I had just finished teaching the two semester sequence of Calculus using the Harvard project [3]. I had also seen a preliminary copy of *Precalculus in Context,* by Davis, Moran, and Murphy [1]. Finally, I had the opportunity to observe two precollege teachers[2] pilot a syllabus in our Introduction to College Algebra course, which puts an emphasis on problem solving, writing, and cooperative work. These experiences as well as discussions with colleagues here and elsewhere influenced my approach to the course.

In preparing for this teaching assignment I was faced with several decisions.

- How much time do I spend reviewing basic algebra skills? (My colleagues who have taught this course warned me about the poor algebra skills of many students.)
- How can I best prepare students for success in Calculus using the Harvard materials, which is the next course in the sequence?
- What role, if any, should writing, projects, and group work play in the course?

Before the semester began the instructors in the precalculus course had a planning meeting to exchange ideas and teaching strategies. A faculty colleague and I left the meeting feeling that we needed to try alternative teaching ideas in the course. (Coincidentally this faculty member had also taught first semester calculus from the Harvard materials.) Below I describe the objectives and teaching approach that I adopted for the course.

- Understanding. A major objective of the course was for the students to develop an understanding of the functions in precalculus through the study of familiar models.
- Writing. The student's work had to be accompanied by a written explanation or justification.
- Projects. Special projects (labs) were assigned every two to three weeks. These were to be written,

[1]Our Fall 1992 freshman class had the following placement distribution: 3% for honors calculus, 28% for calculus, 3% for precalculus, business calculus or finite math, 21% for college algebra, 31% for introduction to college algebra, and 14% unprepared for our lowest level algebra course.

[2]Linda Buckley Griffin, Esperero Middle School, and Lee Jessen, Flowing Wells High School, spent the 1991–92 academic year at the University of Arizona as participants in the University–School Cooperative Teacher Exchange program.

with the mathematical principles highlighted. Conclusions had to be supported by graphs, data, computational examples, and the like. (Most of the labs are from the book by Davis, Moran, and Murphy [1].)

- Group Work. The special projects were to be done by a group of students, three or four to a group. One class period every two weeks was devoted exclusively to work on the labs. In addition, an extra hour was added to class time once every week strictly for lab work. (This was not on the class schedule, and I had a 50% attendance rate.[3]) Students were encouraged to meet outside of class to help each other with the labs. However students had to write their own lab reports, summarizing their own understanding in their own words. (This is a deviation from the lab procedures by Davis, Moran, and Murphy.)

- Technology. The use of all forms of technology, with appropriate explanation, was encouraged. Most students had a TI-81 graphing calculator. They were also encouraged to use the computer software in our undergraduate computer lab.

- Algebra Skills. The course assumed basic algebra and trigonometry skills. The course started with the topic of graphing, found in the third chapter of Holder [2]. However, homework was assigned on the review skills in Chapters One and Two. Class time was also devoted to these skills as they appeared in our work on precalculus concepts. (Those who needed additional help were encouraged to see me or my grader outside of class.)

The main topics covered in the course were functions and graphs, linear and quadratic functions, higher-degree polynomials and rational functions, exponential and logarithmic functions, the trigonometric functions, and polar and parametric equations (Chapters 3–7, 9 of [2]). Although I used a traditional text, the course followed a non-traditional approach in philosophy and teaching style.

Labs

During the semester I assigned five labs. The topics covered were linear behavior, quadratic behavior, graphing exploration, rational functions and asymptotic behavior, and exponential growth. Below is part I of a lab on linear relations and some excerpts from one student's work.

[3]The course has been increased to four credits to reflect a scheduled two-hour lab period added to the course.

Laboratory 1, Part I

Suppose you have a choice of renting a car from two companies—Flat-Rental and Per-Mile-Rental. Flat-Rental charges $40 a day with no charge for miles driven while Per-Mile-Rental charges $25 a day and 20 cents for every mile driven. The object of this lab is to compare the cost of renting a car n days (n fixed) and driving it m miles (m varies). Let's do the analysis in stages.

(a) Suppose $n = 1$ day and you drive m miles (variable). Find the cost function $p(m)$ if you rented from Flat-Rental, $q(m)$ if you rented from Per-Mile-Rental? You can use tables or other means to get started.

(b) Graph $p(m)$ and $q(m)$. [Choose an appropriate viewing rectangle.]

(c) Depending on the number of miles driven, analyze your work in (a) and (b) to determine where you should rent the car and why. Explain.

(d) Do the graphs of $p(m)$ and $q(m)$ intersect? If so, what is the significance of the intersection point?

(e) Repeat (a–d) with $n = 2$ days and m miles.

(f) Repeat (a–d) with $n = 7$ days and m miles.

(g) Can you generalize to a rental of n days (n fixed but unknown) and m miles (variable)? Explain.

Excerpts of One Student's Work on Lab 1, Part I

The first thing I do is write a basic cost equation for each company. I call Flat-Rental "$p(m)$" and Per-Mile Rental "$q(m)$." I let $n = 1$ day of rental and m = miles driven. Flat-Rental costs $40.00 a day so $p(m) = 40n$. Per-Mile Rental costs $25.00 a day plus 20 cents for each mile driven so $q(m) = 25n + .20m$.

The graph of $p(m)$ and $q(m)$ is:

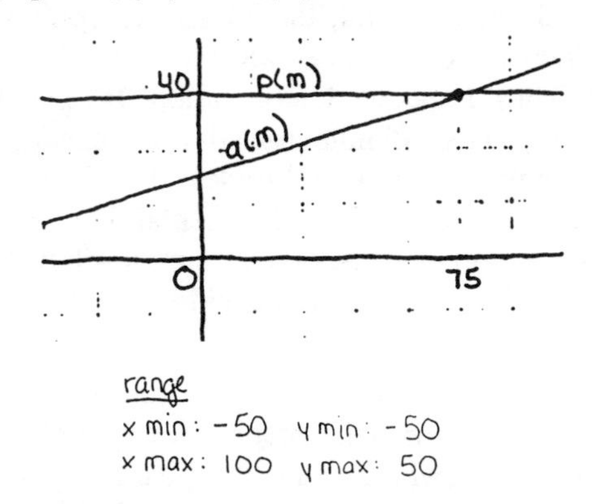

I find that the graphs of $p(m)$ and $q(m)$ intersect at the point $(75, 40)$. This means that if I drive the car exactly 75 miles, the price is $40.00 at both companies. If I drive less than 75 miles in that one day, it is cheaper to rent at

Per-Mile Rental. I know this because the graph of $q(m)$ goes below the graph of $p(m)$, indicating that the cost is less than $p(m)$ from 0 to 75 miles driven. If I drive more than 75 miles, it is cheaper to rent at Flat-Rental. On the graph, this was determined because the graph of $p(m)$ is lower than $q(m)$ when the mileage is greater than 75 for one day.

Again I decide to repeat the steps to compare cost. This time I pick the variable of 7 days and repeat the steps.

	Original equations	If $n = 7$
I)	$p(m) = 40n$	$p(m) = 280$
	$q(m) = 25n + .20m$	$q(m) = 175 + .20m$

II) Graph of $p(m)$ and $q(m)$ if $n = 7$

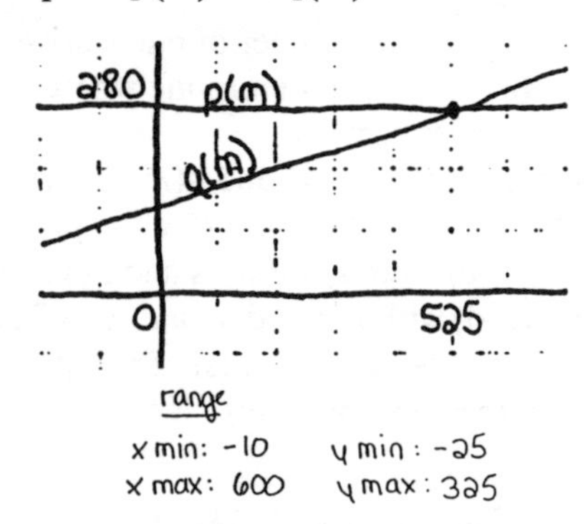

III) The graphs intersect at (525, 280). This tells me that if I drive exactly 525 miles in 7 days, the price will be $280.00 at both companies.

IV) If I drive less than 525 miles, it is cheaper to rent at Per-Mile Rental.

If I drive more than 525 miles, it is cheaper to rent at Flat-Rental.

Finally, I realize that I am seeing a pattern. For each additional day of rental, the intersection points on the graph increase by 75 miles while the cost increases by $40.00. Thus, I make a determination. If I plan to drive over an average of 75 miles each day, it is cheaper to rent at Flat-Rental. If I plan to drive less than 75 miles a day, it is cheaper to rent at Per- Mile Rental. For my leisurely driving around Los Angeles, I am sure it will be less than 75 miles a day. Therefore, I decide to rent at Per-Mile Rental.

From this whole experience of renting a car, I learned many things. I learned that graphs are very useful for visually representing what an equation means. My knowledge of some simple algebraic concepts turned what seemed to be a difficult decision into a rather simple one. The best part is that by taking the time to figure out the math involved in the cost of car rental, I know I will get the better value!

Reactions to the course

I will close with some quotations from student comments taken from anonymous course evaluations given at the end of the semester, followed by my own observations about the course.

Student Comments

The students' comments centered mainly on three areas. First, they expressed concern that the course put a lot of emphasis on the understanding of concepts. Second, they were unhappy with the requirement that a verbal explanation had to accompany the computational steps. Third, they did not feel it was fair that tests included modeling problems that were not identical to ones discussed previously. Below are verbatim quotations from some students.

Understanding of Concepts

"Approaches applications in a logical manner and does not depend on memorization of formulas."

"He carefully explained why we did what we did to make the concepts more easily understood."

Course/Test Expectations

"... the expectations of students in expressing mathematical skills shouldn't be so demanding."

"Different tests, not easier or harder, just different types of questions."

"His enthusiasm seemed to creep into an expectation of his students to feel the same about the course and therefore expected a lot from his students and graded very strictly."

My observations

The course was difficult for practically all the students. Their previous experiences with mathematics prepared them to learn procedures that they were to repeat on homework or tests. They had difficulty transferring the problem solving methods from one setting to a new, but mathematically similar, one. (This was the course's attempt to achieve some of the goals of the paradigm of constructivism.)

As one might expect most students had trouble providing verbal or written reasoning to support their work. There was even some resistance from students who understood the mathematical concepts behind the solution.

I had mixed results on group work. Some of the groups got along well and worked together outside of class; others never became a cohesive unit. I can think of three possible reasons for this. First, the class schedule did not specify that the course had a lab. Second, the 50-minutes of class did not provide sufficient time for the group to progress far enough into the lab. Third, the classroom environment was not conducive to promote group work as the chairs were single-pedestal bolted to the floor.

I was frustrated by the traditional testing format of what David Smith calls "50-minute sound bites" [5]. It did not afford the students time to work nonroutine problems where they had to puzzle through a simple but unfamiliar model.

In spite of various difficulties, I was very satisfied with the progress the students made. Overall I felt that they appreciated that the course tried to explain how mathematical thinking is used to develop "formulas" to solve familiar problems. I had the feeling that this was the first time they were actively engaged in understanding mathematical ideas and how they are developed. The part that troubled them the most was the fact that the tests reflected this approach, which to some meant a lower grade than they would have liked.

Bibliography

1. Davis, M.J., J. F. Moran, and M. E. Murphy, *Precalculus in Context,* PWS-Kent, 1993.

2. Holder, L. I., *A Primer for Calculus,* 5th ed., Wadsworth Publishing Co., 1990.

3. Hughes-Hallett, D., A. Gleason, et. al., *Calculus*, Wiley, 1994.

4. North Carolina School of Science and Mathematics, *Contemporary Precalculus Through Applications,* Janson, 1991.

5. Smith, D.A., "Trends in Calculus Reform," this publication.

Pre-Which-Calculus?—A View From Down Under

Lee E. Yunker
West Chicago Community High School

For secondary teachers responsible for curriculum decision-making, the task of designing an appropriate set of options before calculus has become increasingly more difficult. In recent years, the difficulty has been enhanced by the fact that more secondary students are "calculus ready" earlier than ever before. Further complicating the issue is the fact that the calculus needed by students varies widely depending on the undergraduate course of studies they plan to pursue. There is little doubt the calculus taught for the past half century needs changing, but will the new initiatives to reform it reduce or exacerbate the problem of establishing an appropriate set of pre-calculus options at the secondary level?

During the 1950s, 1960s, and 1970s the pre-calculus courses taught at the secondary level were dominated by the algebraic content related to continuous mathematics, with an emphasis on function related concepts, as well as analytic geometry. The "new mathematics" of the post-Sputnik era had a brief run during this time, but seemingly had little effect on the calculus being taught. Various books had a strong presence in the secondary pre-calculus curriculum during these years. Two books which exemplify this group are *Principles of Mathematics* [1] and *Modern Introductory Analysis* [4], along with their subsequent revisions. Since the pre-calculus curriculum remained virtually unaffected during this time by the rise and fall of the "new math" and "back-to-basics" movements, the "pipeline and filtering system" for students on the road to calculus saw little, if any, change.

However, in the 1980s, with the powerful emerging influence of technology, mathematics had a new awakening. Mathematics found itself basking in the early morning sunlight of a fresh new day. New topics began peeking their heads up like daffodils in the early spring. Included among these are: discrete mathematics; linear programming; matrices; data analysis; linear and non-linear regression analysis; graph theory; recursion; iteration; chaos; fractals; mathematical modeling of non-linear dynamics; and parametric equations, just to name a few. At the same time, cheap, high-powered personal computers and graphing calculators were readily available. This, coupled with NCTM's release of the *Curriculum and Evaluation Standards* in 1989 [9], has forever changed the face of secondary pre-calculus mathematics.

In the mid-1980s a call for more relevance in mathematical content at the pre-calculus level rang out across the country as well. Secondary teachers were looking for more applications along with the use of technology, while undergraduate programs were requiring a course in finite mathematics along with specifically tailored courses in calculus for the Social Sciences and Business majors. In the secondary curriculum, a "fork in the pre-calculus road" developed, one branch leading to advanced placement calculus programs and calculus for the physical sciences, the other leading to calculus for the biological and social sciences and business. The second of these branches takes on a strong discrete mathematics focus, while both rely heavily on contemporary technologies. Some textbooks that represent this changing emphasis include: *Precalculus With Trigonometry* [5]; *Precalculus Mathematics, A Graphing Approach* [3]; *Advanced Mathematical Concepts* [8]; UCSMP *Precalculus and Discrete Mathematics* [7]; *Advanced Mathematics, Precalculus with Discrete Mathematics and Data Analysis* [2]; and *Advanced Mathematical Concepts, Precalculus With Applications* [6].

This brings us to the whole point of this paper. The dilemma facing secondary teachers today is clearly "pre-which-calculus$_{n-1}$?" Most of us, speaking for secondary mathematics teachers, want to develop a set of pre-calculus options which embody the "pump, not a filter" mentality of the 90s, and at the same time adequately prepare our students for the calculus they are about to study. Now couple this with the contemporary calculus reforms we are to discuss at Allerton House, and one has an enormous task to complete. You might say its almost like trying to hit a moving target while on the run yourself. This brings us back to, recursively speaking, "pre-which-calculus$_n$?"

Collegiate mathematics must no longer ignore those of us teaching at the secondary level. Organizations such as the MAA and others, which represent the best interests of undergraduate mathematics, must cooperatively establish guidelines for the prerequisite mathematical skills, instructional approaches, and technologies necessary for success in these various new forms of calculus. Without such guidelines, the transition from secondary mathematics to undergraduate calculus will become even bumpier and more misguided, and an unnecessary national tragedy!

References

1. Allendoerfer, C., and C. Oakley, [1955]. *Principles of Mathematics.* New York: McGraw-Hill.

2. Brown, R. [1992]. *Advanced Mathematics, Precalculus With Discrete Mathematics and Data Analysis.* Boston: Houghton Mifflin.

3. Demana, F., and B. Waits, [1990]. *Precalculus Mathematics, A Graphing Approach.* Menlo Park: Addison-Wesley.

4. Dolciani, M., R. Sorgenfrey, J. Graham, and D. Myers, [1984]. *Modern Introductory Analysis.* Boston: Houghton Mifflin.

5. Foerster, P. [1986]. *Precalculus With Trigonometry.* Menlo Park: Addison-Wesley.

6. Gordon, B., L. Yunker, G. Vannata, and F. Crosswhite, [1994]. *Advanced Mathematical Concepts, Precalculus With Applications.* Columbus: Glencoe.

7. Peressini, A., S. Epp, K. Hollowell, S. Brown, W. Ellis, J. McConnell, J. Sorteberg, D. Thompson, D. Aksoy, G. Birky, G. McRill, and Z. Usiskin, [1992]. *UCSMP Precalculus and Discrete Mathematics.* Glenview: Scott-Foresman.

8. Yunker, L., F. Crosswhite, V. Elswick, and G. Vannata, [1991]. *Advanced Mathematical Concepts.* Columbus: Merrill.

9. National Council of Teachers of Mathematics. [1989]. *Curriculum and Evaluation Standards for School Mathematics.* Reston, VA.

A Publishing Perspective

Publishers, Innovation, and Technology

Peter L. Renz
Academic Press*

Overview

Calculations show that **publishers** do not have the resources to fund educational innovation, moreover it is not their business to do so. This is the business of educational institutions, which have the interest and resources for it. **Innovation** must be a continual process because it is driven by changing societal needs, technology, and the progress of knowledge. The more sharply innovative efforts are focussed, the more likely they are to succeed. **Technology** such as desk top publishing, computer interactive texts, and the like must use what is known about design, type, and illustration if they are to avoid disasters. Developers of new materials should learn as much as they can from publishers and book designers if they want to succeed.

Publishers

Publishing is a business dealing with the transmission and dissemination of ideas. It is, first and last, a business. Publishers must survive and prosper; they must compete. They are neither sources nor the ultimate consumers of the ideas and information in which they deal.

The best of them have consistent editorial direction and mastery of the means of production; they are efficient in promotion, marketing, and fulfillment. They know how to identify and reach markets. They are willing and able to make the most of the material that comes to them. This means being able to tell just what a project is and how to reach its best audience.

A good publisher can tell a sow's ear from a silk purse, and knows how to profit from either. But because of specializations in the business, it is unlikely that a single publisher would be ideal for projects as diverse as, say, a Stephen King novel and an introductory calculus text. Still, a publisher should understand what is needed to succeed in such diverse areas. It may seem that a diversified publishing house such as HBJ (now Harcourt Brace and Company) could handle any sort of project. But I see such houses as a collection of specialized enterprises working together. Giants such as these have their own managerial and financial problems. For HBJ these led to its acquisition by General Cinema, partly in consequence of maneuvers by William Jovanovich to avoid a takeover by Robert Maxwell. Size alone doesn't guarantee stability, nor is it proof against folly.

Publisher's have neither the resources nor skills to develop innovative curricular materials, nor is it their business to do so—essential though they are for dissemination. To make this clear, I begin by looking at the income of college publishing and that of higher education itself. Total college textbook revenues for academic year 1988–1989 were $965.6 million, mathematics being $72.6 million of this total. These figures are from the American Association of Publishers and cover members of that reporting group—not universal for all college publishers, but wide coverage of the industry. Substantial, but the net funds revenues for institutions of higher learning for the year 1987 were $100,438 million, as reported in the 1991 *Statistical Abstract of the United States*. Total college textbook revenues are less than 1% of the revenues of colleges and universities. We must look beyond these overall figures to see what resources might be available for developing innovative materials from both the academic and publishing spheres. Back-of-the-envelope estimates of these sums follow.

For college mathematics publishers the relevant question is: What fraction of textbook revenues might be devoted to development of new courses and materials? For an answer, I turned to Linda Chaput, for many years president of W. H. Freeman and Company. She put the total editorial expense at 8% of net receipts, with developmental editing (close support of an author's efforts with extensive outside reviewing and support of class testing, etc.) being no more than 2.5%. The lion's share of editorial expense goes for acquisitions: the costs of sending editors flying around the country beating the bushes to find and sign tomorrow's projects. We shall see in more detail why predation so dominates cultivation.

Patterns differ from publisher to publisher, but less than you might imagine. Applying Chaput's editorial

*This article was written before the author joined Academic Press.

percentages to the net receipts for mathematics texts gives a total editorial allowance of $5.8 million yearly for college mathematics texts, including $1.8 million for developmental editing. This is a rough measure of publishing's annual budget for acquiring and perfecting mathematics texts. Much of this must be spent polishing existing books or in routine line editing, rather than in supporting more distant and problematic innovations. Further, this $72.6 million dollar annual net is an aggregate figure. No single publisher has anywhere near this income from mathematics texts. Anti-trust laws and competitive instinct make coordination of support impossible, while publishers face the fact that success for innovative materials is low (this is bad) and unpredictable (which is worse), especially early in development.

Taken altogether, this suggests that publishing might be able to make at most something like $1 million per year (about half of the allowance for developmental editing) available for support of innovative texts and materials aimed at college mathematics courses. But, early support of such efforts is risky. Hence, as we will see, the publisher's best strategy is to encourage the players, then, as likely winners appear, make commitments. As a project develops a publisher will be able to better gauge its market and likelihood of success. As the publisher begins to count its chickens before they hatch, but after the chicks are stirring within the egg, it may reasonably encourage and support the author with an advance against anticipated royalties. Royalty rates are roughly 10% to 15% of net, so some of this money is available to buy the time of authors. Note that advances are loans that the publisher expects to recover from the author, either from future royalties or, should the book fail to materialize, by other means. This sobering fact is often overlooked.

The costs and revenues of publishing companies are outlined in my paper, "Style versus Content: Forces Shaping the Evolution of Textbooks and Courseware" [5]. There are systematic variations in the economics of different books depending on the size of the project and its intended market. These are explained and illustrated with numbers drawn from actual cases in the cited paper. I mention here only that overhead runs roughly 50% of net receipts for typical companies. Royalties run roughly 15% of net, while profits after taxes of 10% of net are considered decent for a company of any size.

Note Small, specialized operations such as those of John H. Saxon, Jr. (Grassdale Press), Gilbert Strang, Edward R. Tufte (Graphics Press), Michael Spivak (Publish or Perish), or Bruce Armbruster (University Science Books) operate under special conditions. Initially, a staff of one suffices; there is little overhead. If operations expand, staff is added, and the economics converge toward those of larger houses. These are successful businesses, but only Armbruster, a past colleague from W. H. Freeman and Company, knew what he was getting into. The rest started by having something extraordinary to say and a distrust or dislike of what publishers might make of it.

What of academia's support of curricular innovation? My estimate, checked with such people as Frederick Greenleaf at NYU, Richard Bourgin at Howard University, and Don Albers of Menlo College and the MAA, is that on average 5% of mathematics faculty salary goes for undergraduate course and course materials development. Here I include effort to develop new materials and organize new courses based on notes, selected readings, and the like. I count partial credit for courses which draw on existing books but have substantial new components.

My rough quantification runs thus: I count a total of seven years of teaching at Reed, Wellesley, and Bard colleges in which I developed the equivalent of 6 new semester courses out of a total of about 42 such courses taught. This accounts for just over 14% of my teaching. The strictly innovative part of these courses is a fraction of this 14%. I estimate the fraction was at least 5% of what I was being paid for. Bourgin thought this perhaps a bit high, Greenleaf thought it perhaps low, and Albers thought it close enough for these purposes. Greenleaf estimated his work on innovative course materials as 10% of his time over the last 10 years, but half on graduate courses and half on undergraduate ones. My impressions are that my colleagues, on average, have been at least as active as innovators as I. Moreover, this estimate of at least 5% is consistent with my observations while in graduate school and, very broadly, with what I saw during more than 15 years as an editor and at the MAA.

Using 5% of faculty salaries as a first approximation to the institutional investment in such innovation and noting that there are over 19,000 full-time college and university mathematics faculty members, whose salaries must average at least $35,000 per year, we have a lower bound of $33.25 million per year for academia's investment in innovative courses and materials. I leave out fringe benefits and indirect costs—you know how these run. These resources are in place. The genius of the Alfred P. Sloan Foundation, the National Science Foundation, and their colleagues at the National Research Council and the Mathematical Association of America has been to galvanize and coordinate the troops in systematic efforts to improve teaching of calculus, precalculus, and so on. These efforts have an impact much greater than the direct investment through effects on the enormously greater amounts already available for curricular and course development.

Only colleges and universities are in a position to support generally the development of new course materials. Moreover, they have the right motivation—classroom success, happy students, and pleased future employers of those students. There are many such institutions, and many sorts of approaches can be tried; some will succeed. The granting agencies and foundations can give a sense of direction to all of this by sponsoring conferences and a few select projects with high visibility and a great chance of success. All this will speed the work and widen the impact of successful approaches.

The money for development of curricular materials is simply not to be found in the publishing world. Nor would it be a good thing for a publisher to be trying to force its own project into the classroom. The larger the company's stake in a project, the greater its temptation to win adoptions by fair means or foul. This is inescapable for existing major books that command a big part of the market. The examples *par excellence* are the calculus behemoths. This is where publishers spend their developmental money, in perfecting and enhancing their already successful products. The big money needed to support a large company can be gotten most easily from large projects, ones requiring such monstrous investments and large press runs that only the largest companies can afford to get into the game. Innovation begins at the other end of the scale.

Note that on the whole writing new texts is a losing business. As the above estimates show, royalties in aggregate can never come close to paying back the investment figured as a fraction of faculty time spent toiling in these fields. Fortunately, these efforts serve purposes beyond the writing of materials that earn royalties. Working up new courses and materials helps keep us fresh. It yields new insights for the scholar/teacher, it adds zest to the classroom, it drives departments and schools to reexamine what they are doing and to commit themselves anew to the educational effort.

Given that there is a vast enterprise striving to develop new and exciting materials, publishers should try to pick the best of what is available and make the most of that rather than support specific projects early on. This is why they spend more money on seeking out projects than on developing them. Look to a publisher for a clear, objective view of the chances for widespread adoption of your manuscript, not for a sensitive appreciation of its higher worth. A good editor can give both, but as a business partner you (and his or her employer) must value the former.

Why would an author want such a view? The answer is that the real work of producing a text may well begin after a successful course has been developed. This effort usually goes well beyond what teachers are paid for. At this point, the efforts whose costs I have quantified above are sunk costs; the author can forget them. The question is whether to put more into this effort.

To illustrate, Greenleaf at NYU is several years into development of a multi-course sequence teaching quantitative and scientific literacy. He estimates that work on the manuscript for quantitative literacy now accounts for 50% of his normal effort. He finds a like effort is required for other teaching and administrative duties, and he proposes that through overtime he squeezes in roughly 40% more for research. It adds up to 140%, but we understand this situation. Now he asks, "Is this worth it?" Here the objectivity of an editor is helpful. You can't view your own brainchild objectively, but a good editor can. Greenleaf has published several books and understands the effort needed and the likely rewards. He presses on.

What is the true measure of the broad efforts toward educational innovation? The $33.25 million per year I estimated for the investment being made by four-year institutions is conservative for part of the effort. Consider that many important books have been written by those at two-year colleges or even high schools. I think of the late Charles Miller and his coauthors at American River College and Harold Jacobs of Grant High School, to mention those I know best. There is a large, important enterprise here. Like an iceberg, it is 90% submerged and invisible. Yet without the invisible part, there would be nothing. This is my answer to those who say that one should redirect support from the many projects that will not eventually bear fruit and direct all our efforts toward the winners—chosen somehow before the race is run. This is a mistaken view proceeding from the idea that we are omniscient. We can not tell with certainty what will succeed until we have tried it. More importantly, we can not be sure what direction is best ahead of time, but must observe experiments in many (even, to us, inconceivable) directions, before settling on which way to go.

Innovation

Change is a mixed blessing. Bryan Appleyard's book *Understanding the Present* holds that all our troubles began when Galileo turned his telescope to the moon and observed that it was imperfect, like the earth. Thus vanished the distinction between the celestial and sublunary spheres. This gone, Appleyard argues, the certainty of Thomist thought collapsed and with it went the notion of revealed truth (assured by God and the perfection of St. Thomas's reasoning). I don't believe Appleyard's assertion that mankind felt a comfortable certainty about the universe and his place in it prior to Galileo. The self-consistent, closed nature of the Thomist system that Appleyard cites is not persuasive. How general would he have us suppose a solid grasp of St. Thomas's system was in those days? There may have been a Thomist cast to the thinking then, but my bet is that most people were taken up with the concerns of daily living, not Thomist theology. The advantage of lost paradises is that they can't be examined and found wanting. But Appleyard is right that the movement of science and technology and the shifts in commerce and politics drive current anxiety. This is why nostalgia sells and why people like Appleyard want the lid screwed down once and for all.

The lid will not be screwed down. No final resolution of educational problems will ever be made because, even in mathematics, we must consider changes in our students' needs and backgrounds. Developing new and more appropriate courses and materials must be an endless task. The central issue must be what we ask of a mathematical education. Successful innovation depends critically on a sharp focus on a solvable problem. Here we can learn something from the Devil, in the person of John H. Saxon, Jr.

While the National Council of Teachers of Mathematics (NCTM) was setting forth *An Agenda for Action: Recommendations for School Mathematics of the 1980s,* [4], Saxon Was launching his *Algebra I* [6]. Topping the NCTM agenda is the seemingly laudable recommendation that "problem solving be the focus of school mathematics in the 1980s." A difficulty, made clear in the cited document, is the lack of a definition of "problem solving" or of good ways to measure this ability. Despite the good intent of the framers of this document, this must lead to trouble.

Saxon unerringly went after the softness arising from this lack of clarity. He found that his college students lacked basic skills—manipulative skills, algebra skills. He was irritated by education courses that suggested that one could learn "how to think" without learning content and skills, material that is sometimes dismissed as "rote learning."

Saxon made a limited list of manipulative skills that he thought his students should have mastered and, after checking his list with local high school teachers, he developed his book, which teaches these skills in a gentle way, with built-in spiral review. Saxon's draft material was class tested by teachers who understood and agreed with what Saxon wanted to teach. Saxon, with the teachers, made tests to check student's mastery of the material, and they compared the results on these tests with those of students using other established texts. The results can be guessed from the title of a report of this experiment written by one of the participant teachers, Mickey Yarberry, and published in the *Oklahoma School Board Journal,*: "A Mathematical Dream Comes True" [7]. Read it and weep.

The superiority of Saxon's material is evident. Never mind that Saxon succeeds by setting narrower and far more easily achievable goals. Never mind that the tests, despite the fact that Saxon did not write them himself, are directly keyed to his materials.

Saxon succeeds in large part because he sets for himself and for his teachers and students relatively easily solvable problems. The NCTM's well thought out and well intentioned program founders in part because it sets goals that aren't easy to define or achieve.

I am indebted to Zalman Usiskin for telling me of other tests of Saxon's materials using standard testing instruments in which the advantage, if any, of Saxon's materials is far less evident. (See [1] and [2].) The doubts that arise from such outside examination simply underscore how important it is to a would-be reformer to have sharply defined goals that can be measured to the reformers advantage.

The first requirement of successful educational innovation is to ask the right sort of questions, to set yourself and your students demonstrably achievable (even modest) goals. Loren Larson did a very nice job of this in the table of contents of his *Algebra and Trigonometry Refresher for Calculus Students,* [3]. Its Table of Contents is a self-diagnostic test that checks student mastery of various skills needed in calculus. Larson's book was his answer to the same type of teaching problems that drove Saxon to write, but Larson sought to give the student just a bit of help, rather than setting out to solve the problem by a wholesale transformation of high school mathematics, which seems to be Saxon's goal.

Larson, like Saxon, consulted with local high school teachers to see how things had gone awry with students going on to college. But Larson saw a different sort of need and fashioned a very different answer, less confrontational and more constructive.

Technology

The widespread availability of TEX and other desktop publishing systems has given us "unbelievably badly designed books" in the words of Michael Spivak, originator of AMSTEX, TEX master, and publisher. The power to set justified text creates the illusion of being an expert typesetter. But the standard 8.5 by 11 inch page lays a trap. Run 10 point type from edge to edge and it is very difficult to read. This is why double columns were invented (by Gutenberg).

Placing figures, designing tables, locating relevant illustrative photographs, using design elements to order the material and make it more intelligible, and editing to make a text more understandable, these are all arts that go into bookmaking. The possibilities for various printing and binding methods must be known if you want to get a good product at a reasonable price. Marketing, sales, and fulfillment are not easily done. For every aspect of publishing special skills are helpful.

It must be a given that more and more materials will be stored in computers and published by desk-top methods. If we are to have a continual stream of innovative materials, and this seems needed, then more of the initial work must be done by academics. My plea is that those academics consult with designers, editors, and others who know the requirements and potential of existing technology, so that these new materials will be as attractive as possible.

References

1. Johnson, Dale M. and Blaine Smith, "An Evaluation of Saxon's Algebra text," *Journal of Educational Research,* November/December 1978.

2. Klingele, William E. and Beverly Woods Reed, "An Examination of an Incremental Approach to Mathematics," *Phi Delta Kappan,* June 1984.

3. Larson, Loren, *Algebra and Trigonometry Refresher for Calculus Students,* W. H. Freeman and Company, 1979.

4. National Council of Teachers of Mathematics, *An Agenda for Action: Recommendations for School Mathematics of the 1980s,* NCTM, Reston, VA (1980).

5. Renz, Peter L., "Style versus Content: Forces Shaping the Evolution of Textbooks and Courseware," *New Directions in Two Year College Mathematics,* Donald J. ALbers, ed., Springer-Verlag, New York (1985). Reprinted in *Toward a Lean and Lively Calculus,* Ronald G. Douglas, ed., MAA Notes **6** (1986).

6. Saxon, John H., Jr., *Algebra I,* Grassdale Publishers, Inc., Norman, OK (1981).

7. Yarberry, Mickey, "A Mathematical Dream Comes True," *Oklahoma School Board Journal,* August, 1981.

Part 4

Project Descriptions

The Computer and Graphing Calculator Intensive Calculus Project (C^3E)

Contemporary Precalculus Through Applications (CPTA)

A Core Approach to Calculus

Development of course materials to integrate precalculus review with the first course in Calculus

Earth Algebra

Interactive Mathematics Program (IMP)

Merit Workshop Program in Calculus

NCSSM Contemporary Calculus Project

The Ohio State Computer and Calculator PreCalculus (C^2PC) Project

The Precalculus Revitalization Project

Project CALC: Calculus as a Laboratory Course

Projects and Themes: Calculus at New Mexico State University

The Sensible Calculus Program: Articulating Precalculus and Calculus

The SIMMS (Systemic Initiative for Montana Mathematics and Science) Project

Teacher Enhancement Through Student Research Projects

University of Chicago School Mathematics Project (UCSMP)

The Washington Center Calculus Dissemination Project

The Computer and Graphing Calculator Intensive Calculus Project (C^3E)

OBJECTIVES

A. Use computers and graphing calculators to enhance student understanding of calculus concepts.

B. Increase student's perception of the "value" of calculus through an applications approach.

C. Promote new teaching and learning methods including cooperative learning and inquiry based instruction.

LOCATION

The Ohio State University

CONTACTS

Professors Frank Demana and Bert Waits
Department of Mathematics
The Ohio State University
231 West 18th Avenue
Columbus, OH 43210
Phone: (614) 292-1934
FAX: (614) 292-0694

OTHER SITES

The C^3E calculus project materials (preliminary version) are being used in over 50 high schools and in over 10 colleges and universities throughout the United States.

DESCRIPTION

The C^3E project consists of three interrelated on going activities: (i) curriculum development, (ii) teacher training, and (iii) teacher networking.

Curriculum development The C^3E project materials were developed, piloted, and field tested in both high schools and colleges during the two-year period 1992-1994. The materials will be reflected in a textbook published by Addison-Wesley, *Calculus, A Graphing Approach,* by Finney, Thomas, Demana, and Waits, First Edition, 1994.

Teacher training We have conducted several one-week C^3E summer institutes for high school mathematics teachers since 1991. In summer of 1993, we will conduct 11 C^3E one-week institutes in 11 states. Similar institutes are planned for summers of 1994 and 1995 in many states. Support for these one-week institutes comes from a variety of funding including local Eisenhower funds, US Department of Education, National Science Foundation grants, and public corporation grants.

Teacher networking We have conducted annual national and regional "rejuvenation" weekend teacher conferences since 1988. These events have culminated in a large annual T^3 conference (Teachers Teaching with Technology) dedicated to high school teachers using graphing calculators. These meetings have proved to be very important in providing teacher support during the period they engage in dramatic change (in both curriculum and in integrating technology in their teaching).

Our C^3E philosophy is summarized by the following description of how we approach the mathematics of calculus using computers and graphing calculators. It is based on the principle of incremental change and the fact that inexpensive graphing calculators (now costing less than $50) are practical for *all* students. Our project *assumes* computer or graphing calculator graphing but does not assume the use of computer algebra.

I. Do analytically (with paper and pencil), then SUPPORT numerically and graphically (with a graphing calculator).

II. Do numerically and graphically (with a graphing calculator), then CONFIRM analytically (with pencil and paper).

III. Do numerically and/or graphically, because other methods are IMPRACTICAL or IMPOSSIBLE!

Contemporary Precalculus Through Applications (CPTA)

OBJECTIVES

A primary goal of Contemporary Precalculus Through Applications is to provide an applications-oriented, investigative mathematics course in which students are provided keys to understanding the technological world in which they live. Consistent with statements made in the NCTM *Standards*, CPTA was developed with the belief that:

- graphing calculators should be available to all students at all times;
- a computer should be available in every classroom for demonstration purposes; and
- students should learn to use the computer and graphing calculator as tools for processing information and performing calculations to investigate and solve problems.

The course lays the foundation to support future course work in calculus and other mathematics courses, including finite mathematics, discrete mathematics, and statistics. It also introduces mathematics used in engineering, physical and life sciences, business and finance. In the context of real-world applications and data analysis, students build an understanding of exponential, logarithmic, trigonometric, polynomial, and rational functions.

LOCATION

North Carolina School of Science and Mathematics
Durham, NC

CONTACT

Jo Ann Lutz
NCSSM Mathematics Department
P.O. Box 2418
Durham, NC 27715
(919) 286-3366 ext. 301
e-mail: lutz@odie.ncssm.edu

OTHER SITES

A textbook that has resulted from this project has been published by Janson Publications, Dedham, MA 02026, phone: 1-800-322-6284. The course is now being taught in high schools and colleges across the country, including Eisenhower High School, Yakima, WA; Webb School, Knoxville, TN; Air Force Academy; Trinity College, Hartford, CT; University of Michigan, Ann Arbor.

DESCRIPTION

While preparing students for calculus, Contemporary Precalculus Through Applications offers them the opportunity to look at mathematics more broadly. The course provides a beginning knowledge of real-world applications in the areas of mathematical modeling, data analysis, discrete mathematics, and numerical algorithms. The syllabus lays a foundation to support future course work in the areas of finite and discrete mathematics, statistics, computer programming, and finance, as well as calculus, engineering, and the sciences. Throughout the course, an attempt is made to include non-routine problems from a variety of disciplines. There are models from banking and finance, political science, anthropology, economics, sociology, sports, and international relations, as well as from the sciences.

Six themes spiral throughout the CPTA course: mathematical modeling, computer and calculator as tools, applications of functions, data analysis, discrete phenomena, and numerical algorithms. These themes, which are treated with increasing depth and extension at each exposure, address both the need to prepare for future course work in mathematics and the need to become familiar with the mathematics required for effective citizenship.

The CPTA course is designed to encourage students to approach mathematics in new and innovative ways. A conscious effort is made to combine several themes into examples and problems, ask familiar questions in new contexts, and to apply new concepts to familiar questions. Each theme that spirals through the course has an effect on the instructional approach. Teachers find that having students work with real data, explore the answers to "What if . . . ?" questions, and learn new mathematics in order to answer interesting questions contribute invaluably to the learning process. By seeing immediate usefulness for the mathematics they learn, students develop a real sense of the power of mathematics while learning skills and techniques essential to mathematical investigation in future work.

The computer and graphing calculator are used extensively in the CPTA course. The use of technology expands students' abilities to investigate mathematical concepts, to explore questions, to make conjectures and check answers, to analyze real data, and to discover more fully the mathematical power the course aims for.

CPTA was written because the math faculty at NCSSM felt that their goals could not "be met simply by altering current texts." Now as a textbook, the course is available to other schools so that more students will be exposed to the kinds of experiences that foster the ability to reason mathematically, to develop mathematical power, to understand and appreciate the role of mathematics in contemporary society, and to become confident in their own ability to solve complex problems.

A Core Approach to Calculus

OBJECTIVES

All students at the U.S. Military Academy take a four semester introductory mathematics program. The program begins with a course in discrete dynamical systems, followed by two semesters of calculus, and concludes with a semester of probability and statistics. Non-content objectives of this program include:

- Develop independent learners.
- Develop a modeling approach to problem solving.
- Develop written and verbal communication skills.
- Develop an exploratory approach to learning mathematics.

LOCATION

United States Military Academy

CONTACT

Donald Small
Department of Mathematical Sciences
USMA
West Point, NY 10996

OTHER SITES

Texts used are D. B. Small and J. M. Hosack, *Calculus An Integrated Approach,* McGraw-Hill, 1990, D. B. Small and J. M. Hosack, *Explorations in Calculus with a Computer Algebra System,* McGraw-Hill, 1991, and F. Giordano and M. Weir, *Differential Equations,* Addison-Wesley, 1991.

DESCRIPTION

The discrete dynamical systems course engages students in modeling problems that they understand to be important in their lives (e.g., car payments, investments). The basic questions in dynamical systems (i.e., recursive sequences) concerning long-term behavior provide a natural introduction to the calculus developed through a sequence approach to limits. The calculus course involves several complementary learning activities—computer laboratories, group projects, Gateway testing, and a *core approach* that integrates the treatment of one and several variables. Although the content is traditional including multivariable calculus and differential equations through variation of parameters, the arrangement of topics, approach and emphasis are not. Major emphasis is placed on

- Learning how to learn (mathematics)
- Conceptual understanding
- Learning how to analyze functions (graphically, numerically, symbolically)
- Approximation and error bound analysis
- Understanding and appreciating the generalization process
- Appropriate use of technology.

The *core approach* stresses how to develop mathematical concepts and applications from a minimal core of objects. This, in conjunction with the integrated approach, provides a natural setting for emphasizing exploration, generalization, and discovery. The integrated approach and a strong emphasis on the following two processes provide a uniformity and a streamlining to the course that allows it to fit nicely into two semesters (approx. 100 lessons).

1. The Basic Approximation Process (approximate—improve approximation—generate sequence of approximations—take limit) is central to the development of every major concept.
2. The 5 stage structure for developing all major concepts: (motivation—formulate definition—apply definition to basic functions—develop an algebra—applications)

Group projects and exploratory computer labs actively engage students in *doing* mathematics as well as encouraging collaboration. Project reports and class presentations are used to address the communication objective.

Gateway tests (mastery tests) in graphing, differentiation, and integration provide assurance that students master the basic aspects in these areas. For students who have studied calculus in high school, the integration of single and multivariable functions in the development of the major concepts presents the student with a course that is fresh and clearly different from his or her high school calculus course, thus eliminating the negative attitudes and mediocre results that often result from students who believe that they are just repeating their previous year's course.

Development of course materials to integrate precalculus review with the first course in Calculus

OBJECTIVES

To produce and class-test course materials to accompany a slower-paced first course in calculus that integrates the review of algebra, functions and graphing, and problem-solving strategies with the introduction of the new concepts and techniques of calculus. Our goals are to increase student understanding, improve morale, and attain a higher rate of completion of Calculus I by students who are not prepared to take the traditional one-semester course.

LOCATION

Moravian College, Bethlehem, PA 18018-6650

CONTACT

Doris Schattschneider, Project Director
Department of Mathematics
Moravian College
1200 Main St.
Bethlehem, PA 18018-6650
Phone: (610) 861-1373
e-mail: schattdo@moravian.edu

OTHER SITES

Northampton Community College, Amherst College, Bates College, Lehigh University, Mt. Holyoke College, Randolph Macon College, St. Olaf College

DESCRIPTION

There is a serious problem for those who teach calculus: roughly half of the students with this requirement are inadequately prepared to take calculus. The standard response of most institutions is to offer a one-semester precalculus course and require such underprepared students to successfully complete this course before they are admitted to a calculus course. Until 1988, Moravian College followed this standard pattern. The morale in the precalculus course was low; the attrition in the course was fairly high; the retention of material needed for Calculus I was poor.

In 1988, a new one-year course was introduced at Moravian that replaced the Precalculus-Calculus I sequence. The course integrates the review of precalculus topics *as they are needed* within the first calculus course. In the short time it has been offered, we have observed improvement in morale, level of understanding, and numbers of students completing Calculus I.

Our 2-year FIPSE-funded project has been to prepare, desk-top publish, and class-test supplemental materials especially designed for the slower-paced one year Calculus I course. Our text, entitled *A Companion to Calculus,* is designed to be used along with a standard calculus text and provides review of algebra skills, functions, graphing, and problem-solving techniques in the context of the calculus course.

Evaluation of several aspects of the project has been ongoing: records on student enrollment and completion of the course; comparison of individual student performance on pretests and post-tests; comparison of performance on selected exam questions of students in Calculus I with Review with those in the "regular" one-semester Calculus I course; attitude surveys of those in the integrated course; evaluation of the materials by students and instructors.

A special dissemination conference on the topic of integration of precalculus review in the first course in calculus was held at Moravian College on June 18–19, 1993. Brooks/Cole is the publisher of *A Companion to Calculus,* available June 1994.

Earth Algebra

OBJECTIVES

Earth Algebra is an entry level college algebra course which incorporates the spirit of the NCTM *Curriculum and Evaluation Standards for School Mathematics* at the college level. The context of the course places mathematics in the center of one of the major current concerns of the world, i.e., environmental issues and global warming. There is increased emphasis on conservation and problems of the environment, so most beginning college students enter with at least a superficial knowledge of these issues. Through mathematical analysis of real data, students gain a new perspective on mathematics as a tool. The course incorporates group work, written reports, mathematical models, and the use of technology. All these aspects remove the traditional lecture style as the means of delivery in the course, and make mathematics a hands-on subject.

LOCATION

Kennesaw State College, Marietta, GA.

CONTACTS

Christopher Schaufele and Nancy Zumoff
Kennesaw State College
Mathematics Department
P. O. Box 444
Marietta, GA 30061-0444

OTHER SITES

Earth Algebra was piloted and is now fully integrated into the curriculum at Kennesaw State College. It was tested at selected institutions around the country before publication of the text, *Earth Algebra: College Algebra with Applications to Environmental Issues,* by HarperCollins in January, 1993. As of April, 1994, the text has been adopted by over twenty institutions.

DESCRIPTION

Earth Algebra begins with a class discussion of environmental issues, e.g., global warming, its causes and effects. Relevant real data is available to the class and modeled through class participation and discussion led by the instructor. After initially demonstrating modeling procedures, the role of the instructor evolves into that of providing direction, motivation and serving as a guide to the student. The class is divided into small groups, and much of the work is done in these groups. Communication skills, both written and oral, are required for the group presentations and reports. Some student research of data can be required. The capabilities of modern graphing calculators provide the technology to analyze real data without the tedium and drudgery of repetitive computation and manipulation. Throughout, the practicality of mathematics is emphasized through problem solving, interpretation of results and decision making based on mathematical models. Student of Earth Algebra learn that mathematics can be used as a decision making tool and understand the relevance of mathematics in modern society.

The text is a case study of CO_2 emission and its effect on global warming. Mathematical models are developed from real data with extensive explanations and details given. This provides examples of algebraic techniques, use of the calculator, and the role of mathematics in decision making. Emphasis is placed on the mathematical model as a tool. All the student work is related to a single environmental issue, and all results obtained are used in analysis of this issue and its contribution to the overall problem. Prerequisite algebra sections are placed at the beginning of each part, as needed.

The topics covered in Earth Algebra are: functions; linear, quadratic, exponential, and logarithmic equations and functions; systems of linear equations and inequalities; matrices; geometric series; and linear programming. The course can be used as a "Liberal Arts Mathematics" course or can replace traditional college algebra. It can be taken by all beginning college students with a minimal high school background of Algebra II. Students enjoy an experience in mathematics not available anywhere else in a traditional undergraduate liberal arts program.

Interactive Mathematics Program (IMP)

OBJECTIVES

To create a four-year secondary core curriculum program which will embody the vision of the NCTM *Standards.*

LOCATION

San Francisco State University and the Lawrence Hall of Science at the University of California, Berkeley.

CONTACT

Linda Witnov
Interactive Mathematics Program
6400 Hollis, Suite 5
Emeryville, CA 94608
Phone: (510) 658-6400

OTHER SITES

In 1993–94, IMP will be in use in over 50 school in about a dozen states.

DESCRIPTION

During 1989–1992, the Interactive Mathematics Program (IMP) developed a new, three-year high school mathematics curriculum, which was among the first in the country to embody the vision of the NCTM *Standards.* In this first phase of IMP's work, the new curriculum was field-tested, with diverse populations, in a dozen schools in five states.

In 1992, IMP received a five year grant from the National Science Foundation to continue its work. The goals of this continuation project are:

- to implement the curriculum more widely;
- to evaluate the effectiveness of the curriculum for students;
- to assess the needs for teacher in-service in large-scale implementation; and
- to write a fourth year curriculum, while preparing a version of the curriculum for publication.

The project as a whole is under the direction of Lynne Alper and Sherry Fraser, of the EQUALS program at the Lawrence Hall of Science, University of California, Berkeley, and Professors Dan Fendel and Diane Resek, of the Mathematics Department at San Francisco State University.

The IMP program is a major departure from traditional high school mathematics, in terms of curriculum structure and content, and in pedagogy.

While the traditional Algebra I–Geometry–Algebra II structure is designed around individual mathematical skills and concepts, the IMP curriculum is problem-based, consisting of 4-week to 8-week units that are each organized around a central problem or theme. Motivated by this central focus, students solve a variety of smaller problems, both routine and non-routine, that develop the underlying skills and concepts needed to solve the central problem in that unit.

In addition to this emphasis on mathematics in context, the IMP curriculum introduces major changes in mathematical content, including substantial work with probability, statistics, curve fitting, and discrete mathematics. The curriculum also provides numerous opportunities for students to engage in long-term investigations and projects, working both independently and cooperatively.

In IMP's view of the mathematics classroom, the teacher is facilitator and supporter of learning, and not just an imparter of knowledge. Much of the classroom work takes place with students working collaboratively. By fostering the sharing of ideas among group members, such activities establish the notion that communication is an important component of mathematics.

Merit Workshop Program in Calculus

OBJECTIVES

To increase the number of undergraduates in math and science based majors from groups currently underrepresented in these fields. In particular, to increase the number of students from these groups who excel in the beginning calculus courses.

LOCATION

University of Illinois at Urbana-Champaign

CONTACT

Paul R. McCreary
178 Altgeld Hall
Department of Mathematics
University of Illinois at Urbana-Champaign
Phone: (217) 244-1659

DESCRIPTION

Participants The Merit Workshop Program is designed primarily for groups currently underrepresented in the science and mathematics based majors. At this university such groups include students from African-American and Hispanic backgrounds, students form small high schools, and female students. Entering first-year students from these groups who have chosen a math/science based major are invited to participate in the program. Each year approximately 70 students choose to participate in the extra class hours and collaborative group activities in first and second year calculus courses.

Structure Participants in the program attend the same lectures, complete the same homework assignments and take all exams and quizzes together with other students in their large lecture section. The workshop participants spend and additional four hours per week in class sessions working collaboratively on problems significantly more difficult than the homework assignments. This collaborative work is neither graded nor corrected, but is employed to encourage student-to-student interaction and to allow immersion into interesting and open-ended calculus problems. Often, unless an exam is immanent, incomplete and even inaccurate ideas articulated by students in the group sessions are left uncorrected by the teaching assistant in charge. Such inaccuracies have been observed to disappear without intervention by instructors. The students are allowed and encouraged to explore and investigate their unformed ideas. They develop an enhanced intuitive understanding of the material and get better at persevering and making multiple attempts at difficult problems.

Results Over the past three years the very highest exam average in the large lecture section of which the program's two sections are a part has been scored by a workshop participant. This past Fall the top four scores out of 139 belonged to Merit Workshop students. The workshop students, on the average, earn almost 2 grade points higher than students from similar background in the course and almost a full grade point higher than all students in the course. The hour and final exams are all graded in common grading sessions including all teaching assistants and the lecturer to assure uniform grading standards. Further, the participants of the Merit Workshop Program typically have standardized test scores (e.g., ACT math subscores) slightly *below* that of all students from similar backgrounds, i.e., minority or rural backgrounds, and substantially below that of all students taking the calculus courses.

NCSSM Contemporary Calculus Project

OBJECTIVES

This course is designed to address three very broad shortcomings in most existing calculus courses. Our objectives are as follows:

- students should be able to apply calculus to problems that are stated in new and different contexts;
- calculus is applied to virtually every field of human endeavor. Students should appreciate and understand a broad range of these applications, and they should be able to use mathematics as a tool for modeling a variety of real-world phenomena; and
- the numeric, graphic, and symbolic capabilities of computers and calculators should play an important role in the mathematics classroom.

LOCATION

The early work on this project was done in conjunction with Duke University. For the past two years, all of the work on this project has been done at the North Carolina School of Science and Mathematics in Durham, North Carolina.

CONTACT

Jo Ann Lutz or John Goebel
North Carolina School of Science and Mathematics
Box 2418
Durham, NC 27715
Phone: (919) 286-3366
e-mail: lutz@odie.ncssm.edu or
goebel@odie.ncssm.edu

OTHER SITES

Montgomery Blair High School, Silver Spring, MD; Southern Senior High School, Harwook, MD; Riverside High School, Durham, NC.

DESCRIPTION

This course provides students with an applications-oriented, investigative calculus course in which students are provided tools for understanding the world in which they live. The course involves students in both the development of problem statements as well as in their solution. The students learn to use the concepts of calculus to solve problems in a variety of contexts, many of which are discussed over extended periods of time.

The fabric of the course is woven with seven themes that are an indispensable part of virtually every segment of the course. The themes address important student needs identified by concerned professionals. Each of the seven themes is described below.

1. Developing Understanding Calculus offers students a repertoire of new techniques for describing the world around them and should be accessible to as wide an audience as possible. For that reason, this course encourages intuitive understanding of many topics. Mathematical rigor, although appropriate at times, should not be an obstacle to the success of students. Graphical, numerical, and algebraic interpretations are used together whenever possible to improve student understanding. Students will discover concepts for themselves as often as is feasible, often through mathematics laboratory experiences.

2. Applications and Mathematical Models Problems encountered in the course describe realistic situations that can be modeled using calculus. As much as possible, major concepts of the course are introduced through applications and mathematical modeling. These applications come from all areas in which calculus is used as a tool to solve problems. As students see a wide variety of applications, they will begin to appreciate the power of calculus. Translation of a problem from a verbal description to the notation of calculus is of paramount importance in this course.

3. Computer and Calculator as Tools Much of traditional calculus involves extensive paper and pencil manipulation. In this course, electronic sketchpads, computer algebra systems, graphing packages, and scientific word-processing software are all tools that are used to present and discover concepts. Technology is also used to experiment with different values for parameters in problems, to try different strategies for solution, to test out various conjectures, and to ask "What if?" questions that otherwise would not be feasible to investigate. Students must, however, know when a result is reasonable and how to interpret results, requirements that further emphasize the need for students to understand calculus.

4. Numerical Algorithms Approximations for the derivative, the definite integral, and solutions to differential equations can be found using numerical algorithms. Since the techniques used in the algorithms reinforce the broad concepts of calculus, the algorithms provide more than just the solution to a problem. Numerical algorithms do have limitations, and students will develop understanding of the accuracy and the appropriateness of various numerical methods.

5. Discrete Phenomena Calculus involves the study of phenomena that are represented by continuous functions. Fast computing makes possible the study of these phenomena using discrete techniques. Finite differences, difference quotients, and other discrete concepts help us move from the discrete to the continuous domain. Frequent links between discrete and continuous phenomena are emphasized.

6. Computer Laboratory Experiences Computer Lab work is the centerpiece of the entire course. Students use the computer as a tool for investigation and discovery. Lab activities allow students to discover concepts, to develop their intuition related to calculus, and to investigate extended problems in groups and individually. These labs frequently culminate in written reports that summarize the particular investigation and draw conclusions from it.

7. Writing about Mathematics Throughout this course, students will write about mathematics, both the concepts and the interpretations of these concepts. The students will use the language of mathematics and communicate their ideas to other individuals familiar with the subject. Writing about mathematics completes a process that begins with the translation of words about a problem into the language and notation of mathematics, then uses mathematics to investigate the problem, and ends with translating the results back into a verbal explanation and summary.

The Ohio State Computer and Calculator PreCalculus (C^2PC) Project

OBJECTIVES

A. Prepare students for a "modern" calculus.

B. Use graphing calculators to enhance student understanding of functions and related concepts.

C. Increase students' perception of the "value" of mathematics through applications.

D. Promote new teaching and learning methods including cooperative learning and inquiry based instruction.

LOCATION

The Ohio State University

CONTACTS

Professors Frank Demana and Bert Waits
Department of Mathematics
The Ohio State University
231 West 18th Avenue
Columbus, OH 43210
Phone: (614) 292-1934
FAX: (614) 292-0694

OTHER SITES

The C^2PC project materials are now being used in over 1000 high schools and in over 100 colleges and universities in 44 states in the United States.

DESCRIPTION

The C^2PC project consists of three interrelated on-going activities: (i) curriculum development, (ii) teacher training, and (iii) teacher networking.

Curriculum development The C^2PC project materials were developed, piloted, and field tested in both high schools and colleges during a five year period 1983-1988 with support from the Ohio Board of Regents, British Petroleum of Ohio, and the National Science Foundation. The materials are now reflected in two textbooks published by Addison-Wesley.

Precalculus Mathematics, A Graphing Approach, by Demana, Waits, and Clemens, Third Edition, 1994.

College Algebra and Trigonometry, A Graphing Approach, by Demana, Waits, and Clemens, Second Edition, 1992.

Teacher training We have conducted over 50 one-week C^2PC summer institutes for high school mathematics teachers since 1988 in over 17 states. In the summer of 1993, we conducted 26 C^2PC one-week institutes in 18 states. Also similar institutes are planned for summers of 1994 and 1995 in many states. Support for these one-week teacher training institutes comes from a variety of funding sources including local Eisenhower funds, US Department of Education grants, National Science Foundation grant, and public corporation grants.

Teacher networking We have conducted annual national and regional "rejuvenation" weekend teacher conferences since 1988. These events have culminated in a large annual T^3 (Teachers Teaching with Technology) dedicated to high school teachers using graphing calculators. These meetings have proved to be very important in providing teacher support during the period they engage in dramatic change (in both the curriculum they teach and in integrating technology in their teaching).

Our C^2PC philosophy is summarized by the following description of how we approach precalculus mathematics using graphing calculators. It is based on the principle of incremental change and the fact that inexpensive graphing calculators (now costing less than \$50) are now practical for *all* students.

I. Do analytically (with paper and pencil), then SUPPORT numerically and graphically (with a graphing calculator)

II. Do numerically and graphically (with graphing calculator), then CONFIRM analytically (with paper and pencil)

III. Do numerically and/or graphically, because other methods are IMPRACTICAL or IMPOSSIBLE!

The Precalculus Revitalization Project

OBJECTIVES

- To train faculty, mainly from community colleges, in new topics and approaches for teaching precalculus.
- To produce a sourcebook of activities, projects, and instructional strategies.

LOCATION

Seattle University

CONTACT

Carl Swenson
Math Department
Seattle University
Seattle, WA 98122
Phone: (206) 296-5926

OTHER SITES

Seattle Central Community College, Janet Ray

DESCRIPTION

This was a two-year program funded by NSF's Division of Undergraduate Education. A five-day workshop held in June 1992 featured presentations by recognized experts, both national and local. Topics included: use of collaborative and active learning modes, use of graphing calculators and other technology, use of physical experiments and data collection in the classroom, and development of mathematical models from industry.

The workshop format encouraged the sharing of materials and expertise from within the group of faculty participants as well as from guest presenters. The issue of what is appropriate course content, given the changes in calculus, prompted considerable spirited discussion. During the year following this workshop, faculty agreed to teach at least one precalculus course using materials and techniques developed.

In both the planning and implementation phases there was significant collaboration between two and four year schools. The Co-PI's had worked together on the Washington Center Calculus Dissemination Project and saw the need for work at the precalculus level as a necessary next step. The first workshop was attended by 23 faculty representing seven community colleges, two four-year colleges, and four high schools.

A second workshop was held in June 1993 and involved over 20 more faculty. In addition to training as before, participants had an opportunity to review materials developed during the preceding year by the first group of faculty. An edited version of materials developed by participants will be published in the form of a Precalculus Sourcebook.

Project CALC: Calculus as a Laboratory Course

OBJECTIVES

To develop a 3-semester calculus course that will empower students to

- use mathematics to structure their understanding of and to investigate questions in the world around them;
- use calculus to formulate problems, to solve problems, and to communicate the solutions of problems to others;
- use technology as an integral part of this process of formulation, solution, and communication;
- work and learn cooperatively.

LOCATION

Duke University

CONTACT

David Smith and Lawrence Moore
Department of Mathematics
Duke University
Box 90320
Durham, NC 27708-0320
Phone: (919) 660-2825
FAX: (919) 660-2821
e-mail: das@math.duke.edu or
lang@math.duke.edu

OTHER SITES

Albertson College (ID), Alverno College (WI), Big Bend Community College (WA), Bowdoin College (ME), University of California-Irvine, Central Oregon Community College, East Carolina University (NC), Evergreen State College (WA), Frostburg State University (MD), Lynchburg College (VA), Medgar Evers College (NY), Mercer University (GA), Middle Tennessee State University, University of Mississippi, Mt. Holyoke College (MA), Principia College (IL), Riverside High School (NC), Rockhurst College (MO), College of St. Francis (IL), St. Mary's University (TX), Seattle Central Community College (WA), Weber State University (UT), West Virginia University, Widener University (PA).

Description The key features of the Project CALC course are *real-world problems, hands-on activities, discovery learning, writing and revision or writing, teamwork, intelligent use of available tools, and high expectations of students.* Our three-semester calculus program is based on an ideal laboratory science model. Students working in pairs explore real-world problems with real data, conjecture and test their conjectures, discuss their work with each other, and write up their results and conclusions on a technical word processor. This laboratory experience drives the rest of the course. It shapes the contents and the approach of the text and the format of the classroom activities.

In the classroom, we concentrate less on *teaching* and more on *learning*. We now lecture less and devote more time to student activities in groups—e.g., gathering data on periods of pendulums, balancing plywood cutouts to locate centers of mass, deciding whether two approaching planes will collide, and finding parametric representations of curves drawn with a Spirograph.

In most implementations, the laboratory is a computer laboratory, but we are developing a calculator-based version of the labs for HP-48 and TI-85 calculators. Full implementations exist for a Mathcad/Derive combination on DOS machines, for Windows versions of Mathcad (with Maple built-in), and for *Mathematica* on Macintosh, NeXT, and DOS/Windows computers. Other implementations in development include *Maple* and *Derive* (without *Mathcad*).

Project materials (text, lab manuals, instructor's guide, newsletter) are available from D.C. Heath, 125 Spring St., Lexington, MA 02173, (617) 862-6650.

Projects and Themes: Calculus at New Mexico State University

OBJECTIVES

Get students to think for themselves on major multistep, take-home problems, working individually or in groups. Alter fundamentally students' view of what mathematics is all about and simultaneously build their self confidence in what they can achieve through imaginative theoretical thinking.

LOCATION

New Mexico State University

CONTACT

David Pengelley
Department of Mathematical Sciences
NMSU
Las Cruces, NM 88003-0001
Phone: 505-646-3901
e-mail: davidp@nmsu.edu

OTHER SITES

Yes, but extent unknown.

DESCRIPTION

The calculus program for science and engineering students at New Mexico State University has gone through several major evolutionary stages in the past five years, each prompted by input from new groups of faculty. Today, the vast majority of faculty teaching calculus are voluntarily involved in this ongoing process. Our experiences can serve as an example of how widespread change can occur in a department of over thirty research faculty at a state university, teaching calculus sections of forty-five students.

The introduction of two-week student assignments in 1987 led to the NSF funded development of a "student research projects" program. The projects resemble mini-research problems. Most of them require creative thought and all of them engage students' analytic and intuitive faculties, often weaving together ideas from many parts of calculus. While many are couched in a seemingly real-world setting, often with an engaging story line, they are all in a sense theoretical; one cannot do them without an appreciation of the ideas behind the method. Students must decide what the problem is about, what tools from calculus they will use to solve it, find a strategy for its solution, and present their findings in a written report. This approach yields an amazing level of sincere questioning, energetic research, dogged persistence, and conscientious communication from students. Moreover, our own opinion of students' capabilities skyrocketed as they rose to the challenges. Typically, we assign two to three major projects each semester. These projects are completed entirely outside of class and contain material over and above the day-to-day course work. Over one hundred of these projects are published in the book *Student Research Projects in Calculus* (Mathematical Association of America, 1992). This book also has chapters detailing the logistics of assigning projects and advice for instructors.

As new faculty became involved, they injected fresh ideas and the projects approach evolved. In 1991, a collection of faculty pioneered a major new emphasis incorporating cooperative learning and self instruction both in and out of the classroom. Using the pedagogical ideas of projects, they developed structured in-class group assignments called "themes." A distinct change is that themes are used to introduce core material and much class time is spent working on them, with less time for lecture. After spending a couple of semesters experimenting with as many as twelve themes during a fifteen week semester, classes now alternate between a one or two week theme and one or more weeks of more traditional classroom activity, with a total of four to six themes per semester. Another recent experiment is mastery-based skills examinations to ensure command of differentiation and integration skills.

Themes and skills examinations have made possible a shift on midterm and final examinations away from rote calculation towards more conceptual and essay questions. Theme assignments have also prompted us to incorporate means of improving student writing skills in mathematics. By coordinating the creation of themes material for all instructors of a course, we have made the extra demands on the instructors small. We have found the extra work rewarded as we observe a marked increase in pass rates of the students in calculus classes which use themes, as well as in the students' subsequent mathematics courses. Finally, we are discovering something very encouraging: when students are involved in collaborative self learning with themes, we are less pressed to "cover" material in class, so the syllabus feels less rushed. The difference may be substantial.

The Sensible Calculus Program: Articulating Precalculus and Calculus

OBJECTIVES

The Sensible Calculus Program (SCP) addresses the need to improve calculus learning. By redesigning the college precalculus course while continuing to complete work on a sensible first year calculus course, the SCP is developing a precalculus/calculus textbook and program with a consistent, thematic, and conceptual approach. The precalculus component provides motivation, background, concept development, and skills for real learning. The calculus component builds on these foundations with an approach that extends the precalculus maturation into calculus.

LOCATION

Humboldt State University

CONTACT

Professor Martin E. Flashman
Department of Mathematics
Humboldt State University
Arcata, CA 95521
Phone: (707) 826-4950
e-mail: flashman@axe.humboldt.edu

OTHER SITES

Test sites now being sought for 1994–96.

DESCRIPTION

The calculus component of the SCP focuses on the themes of differential equations and estimation throughout the first year of calculus, using modeling as a central motivation for applications of the calculus. An important and unique feature of the SCP is that interpretations of calculus concepts consistently refer to both a geometric/tangent view using the graph of a function and a dynamic/motion view using transformation figures to visualize functions. This approach is particularly helpful for students who are still developing a basis for understanding the function concept.

With an early discussion of differential equations, SCP introduces students first to a visual approach through tangent (direction) fields. Then a combined numeric and visual approach to DE's is developed using Euler's method. These complement the usual symbolic discussions of antidifferentiation and the indefinite integral. The SCP uses models to motivate transcendental functions with differential equations. This approach shows how common functions arise and/or remain significant because they solve problem situations modeled with first and second-order differential equations.

The SCP approaches Taylor theory as a distinct and important part of calculus based on differential equations with estimation issues. The calculus of Taylor polynomials (not series) appears as a tool for approximating difficult definite integrals such as $\int_0^1 e^{-x^2}\,dx$ with a sensible control on the error. Estimating the solution to differential equations provides further motivation for the convergence questions of infinite series.

Probability in the SCP Since 1991, the SCP has been developing new materials on continuous probability starting from a simple experiment of throwing a dart at a unit circle. Trying to understand the distribution function for this random variable eventually leads to its density function as its derivative. The SCP includes probability concepts as regular alternative interpretations of the calculus.

Precalculus Issues The SCP addresses directly the background, motivation and experience of students entering the calculus course. The conceptual approach used in the SCP is prepared in its precalculus component, working on a consistent approach to functions using the same transformation (mapping) figures used in the calculus component as well as graphs and tables of computed numerical values to examine function concepts throughout the course. The SCP calculus materials build on these visual tools along with their interpretations, leading to a more sensible and thematic approach to calculus.

The project plans to introduce several other concepts that should prove useful to students without reference to a future calculus course. Probability concepts will appear in the precalculus course. Many calculus concepts for estimation (such as bisection and Newton's method) will also be introduced in the precalculus component along with historical discussions and applications (including probability).

Technology and the SCP The SCP is not committed to any specific technology but presumes access to technology capable of graphing and computation. Programming features are useful for some computation or visual examples not found in the technology.

The SIMMS Project

OBJECTIVES

The Systemic Initiative for Montana Mathematics and Science (SIMMS) is a cooperative project of the state of Montana and the National Science Foundation. The SIMMS Project, funded through the Montana Council of Teachers of Mathematics, has the following objectives:

- Promote integration in science and mathematics education.
- Redesign the 9–12 mathematics curriculum using an integrated interdisciplinary approach for all students.
- Develop and publish curriculum and assessment materials for grades 9–16.
- Develop a shared vision of K–12 science education.
- Incorporate the use of technology in all facets and at all levels of mathematics education.
- Increase the participation of females and Native Americans in mathematics and science.
- Establish new certification and recertification standards for teachers.
- Redesign teacher preparation programs using an integrated interdisciplinary approach.
- Develop an inservice program on integrated mathematics to prepare teachers of grades 9–16.
- Develop the support structure for legislative action, public information, and general education of the populace necessary for effective implementation of new programs.

CONTACTS

Dr. Johnny Lott
SIMMS Project
Department of Mathematical Sciences
University of Montana
Missoula, MT 59812-2313
Phone: (406) 243-2696

Dr. Maurice Burke
Department of Mathematical Sciences
Montana State University
Bozeman, MT 59717-0240
Phone: (406) 994-5330

DESCRIPTION and LOCATIONS

The SIMMS curriculum, written by secondary teachers for grades 9–12, will interface with the mathematics curriculum being developed at the University of Montana for grades 6–8 by the NSF funded Six Through Eight Mathematics (STEM) Project. The collaboration of mathematics/science teachers and university faculties in Montana is producing a curriculum solidly based in applied contexts meaningful for all students.

The Professional Development component of SIMMS is providing inservice for Montana teachers. In addition, preservice teacher-training programs are being redesigned by this component.

The Assessment Component of SIMMS is gathering information at the student, classroom, curriculum, and state/project levels. In addition, SIMMS is acting as an agent of change in the way assessments are done at all of these levels.

The Montana Legislature has allocated $2,000,000 for technology in secondary schools for the SIMMS Project. Currently 82 schools have been funded with more to be funded in 1994–95. Clearly Montana sees the need and is eager to improve the mathematics education of its students.

The SIMMS Project is working with current initiatives in Native American education. The American Indians Research Opportunities program, the American Indians in Mathematics program, and the Alliance of States Supporting Indians in Science and Technology are providing the SIMMS Project avenues to increase the numbers of Native Americans choosing careers in mathematics and science.

Teacher Enhancement Through Student Research Projects

OBJECTIVES

Share with high school mathematics teachers our use of student research projects in mathematics classes and cooperatively develop a model program demonstrating the viability of the use of projects. Develop high-level problem solving abilities in students and incorporate writing and cooperative learning in mathematics classes.

LOCATION

Department of Mathematical Sciences, New Mexico State University, in conjunction with Las Cruces High School, Mayfield High School and Oñate High School, Las Cruces Public Schools, Las Cruces, NM

CONTACT

Douglas S. Kurtz
Department of Mathematical Sciences
NMSU
Las Cruces, NM 88003-0001
Phone: (505) 646-3901
e-mail: doug@nmsu.edu

DESCRIPTION

Beginning in 1987, mathematicians at New Mexico State University instituted a curriculum development program in calculus classes. The focus of their work was the introduction of two-week student assignments called "student research projects". The projects get students to think for themselves on major multistep, take-home problems, working individually or in groups. They engage students' analytic and intuitive faculties, often weaving together ideas from many parts of calculus and practical experience. Complete solutions are presented in a well written, grammatically correct report. The viability of the university program became evident during several years of development, implementation and evaluation. Some of the mathematicians involved in this wanted to go further and considered incorporating the pedagogy of projects into high school classes.

Serendipitously, some high school mathematics teachers approached the department to seek ways to better prepare their students and we began a cooperative curriculum development program working with them. We told the teachers about our pedagogical methods and gave advice on how to get started. Teachers created the project assignments themselves and experimented with different strategies for using them. The program evolved over a three year period: the first year involved small scale experimentation by a cadre of seven teachers; the second and third years added twenty-five more teachers, with the initial group serving as mentors. Projects have been used in classes ranging from Algebra I to Calculus, and from remedial to honors courses.

The teacher training consisted of summer courses and workshops during the academic year. The summer courses included an introduction to project pedagogy and use, problem solving, project writing, and the discussion of more general issues, such as cooperative learning and equity issues. The workshops were essentially round table discussions during which teachers related their recent experiences; new techniques were discussed and considered for implementation. This open ended forum encouraged teachers to try new ideas in their classrooms and get advice from other teachers on their implementation.

After two years, a model program emerged. Teachers typically assign four to six major projects each year. The initial projects contain material over and above the day-to-day course work, though more recent ones have incorporated core material from the classes. Students work on the projects both in and out of the classroom. Almost all project assignments are designed to be solved by groups of three to four students. This has helped the teachers to incorporate cooperative learning into their classes.

The introduction of projects into the classroom has convinced high school mathematics teachers that their students can learn mathematical material on their own. This allows teachers to spend less time lecturing and gives students more time to spend on learning. It has also been a better way to cover topics which traditionally are difficult for students to learn.

University of Chicago School Mathematics Project (UCSMP)

OBJECTIVES

Since its inception in 1983, UCSMP has focused on significantly upgrading the school mathematics experience of the average student in grades K–12. The secondary component (the only component described here) has developed a full mathematics curriculum for students in grades 7–12 that implements the recommendations of many national groups, and conducts workshops and meetings about these materials.

LOCATION

The UCSMP secondary component's offices are in the Department of Education at the University of Chicago and the component is co-directed by professors of education and mathematics. Each of the six textbooks in the secondary curriculum was written by a team of authors assembled on campus and was tested in schools across the nation.

CONTACT

Carol Siegel
Department of Education—UCSMP
University of Chicago
5835 S. Kimbark
Chicago, IL 60637
Phone: (312) 702-9770
FAX: (312) 702-0248

OTHER SITES

UCSMP secondary materials have been available for several years and have been successfully implemented in schools in all 50 states.

DESCRIPTION

The UCSMP secondary curriculum includes six courses, entitled *Transition Mathematics; Algebra; Geometry; Advanced Algebra; Functions, Statistics, and Trigonometry;* and *Precalculus and Discrete Mathematics.*

Several basic elements distinguish UCSMP materials from standard courses: **wider scope**, with geometry, algebra, and some discrete mathematics in all courses, statistics and probability integrated into the study of algebra and functions, and the inclusion of recent developments in mathematics and mathematical applications; emphasis on **reading and problem-solving**; an abundance of **applications** throughout the materials; use of **technology**, with all students expected to have scientific calculators and do some computer work in all courses and function-graphing technology used in the last three courses; a **multi-dimensional approach to understanding**, emphasizing skills, properties, uses, and representations; and an **instructional format** featuring continual review combined with a modified mastery learning strategy.

The secondary curriculum is intended for any student possessing mathematical knowledge at a 7th grade level regardless of age. Thus a student who begins *Transition Mathematics* at 6th grade can complete the entire sequence and take AP calculus as a senior. A student may begin *Transition Mathematics* as late as 9th grade and still complete a high-school mathematics sequence through *Advanced Algebra*. UCSMP's goal is to have every high-school graduate take the first four courses (through *Advanced Algebra*), all college-bound students take the first five (through *Functions, Statistics, and Trigonometry*), and all students who may study technical subjects take all six courses (through *Precalculus and Discrete Mathematics*).

Evaluation has been integral to UCSMP from the start, providing a regular source of feedback. UCSMP evaluators use the latest quantitative and qualitative methods to assess the impact and implementation of project curricula. Nationwide comparative studies have shown that UCSMP students significantly outperform students in conventional courses on the nontraditional content covered in UCSMP texts while holding their own on traditional content.

The Washington Center Calculus Dissemination Project

OBJECTIVES

- To adapt and disseminate the calculus reform work carried on by both Harvard and Duke.
- To provide training for faculty in active learning modes and the use of technology.

LOCATION

The Evergreen State College

CONTACT

Robert Cole
Lab I
The Evergreen State College
Olympia, WA 98505
Phone: (206) 866-6000

OTHER SITES

Seattle Central Community College, Janet Ray

DESCRIPTION

This was a two-year program funded by the NSF's Division of Undergraduate Education. Workshops held in September 1991 and July 1992 for mathematics faculty from Washington State featured presentations given by principals from the Harvard Calculus Consortium and Duke's Project CALC. In addition, local experts led discussions on uses of collaborative learning, graphing calculators, symbolic algebra systems, experiments and data collection,and student projects.

The activities were organized under the auspices of an existing consortium, The Washington Center for the Improvement of Undergraduate Education. The Center focuses on low-cost, high-yield approaches to educational reform, especially those involving interdisciplinary studies and active learning strategies. Its Winter 1993 newsletter, a nationally circulated document, reported on the project outcomes in detail.

Participants generally came in teams of at least two per institution. Those selected were asked for a commitment to implement what they learned in concrete ways and to document results. Modifying materials so they were appropriate to individual settings was encouraged.

The Co-PI's followed up summer workshops with campus visits designed to address implementation problems and to involve other members of the mathematics department. A two-day workshop was held each spring to bring participants back together for sharing and debriefing.

Twenty-one schools (13 community colleges, 6 four year schools, and 2 high schools) involving about 75 faculty have participated in the project. Elements of the workshop, both content and pedagogy, have been implemented in virtually every participating department. In addition, the calculus activities have spawned major revisions in related courses, including precalculus, multivariable calculus, linear algebra, and differential equations.

Preparing for a New Calculus Conference Participant List

April 22–25, 1993

Donald Albers
Associate Executive Director and
Director of Publications and Programs
Mathematical Association of America
1529 Eighteenth St., NW
Washington, DC 20036

Lyle Anderson
Department of Mathematical Sciences
SIMMS Project
Montana State University
Bozeman, MT 59717

Chris Avery
Department of Mathematics
DeAnza College
21250 Stevens Creek Blvd.
Cupertino, CA 95014

Beverly Black
Center for Research on Teaching and Learning
University of Michigan
109 East Madison
Ann Arbor, MI 48109

Gary Bogar
Department of Mathematical Sciences
Montana State University
Bozeman, MT 59717

Peter Braunfeld
NSF Room 635
Division of Elementary Secondary & Informal Science
4201 Wilson Blvd.
Arlington, VA 22230

Judith Broadwin
Department of Mathematics
Jericho High School
6 Yates Lane
Jericho, NY 11753

Morton Brown
Department of Mathematics
University of Michigan
Ann Arbor, MI 48109

Donald Bushaw
Washington State University
French Admin. Building
Room 422
Pullman, WA 99163

John Choate
Department of Mathematics
Groton School
Box 991
Groton, MA 01450

Jon Cruver
Department of Natural Sciences
Haskell Indian Jr. College
155 Indian Avenue
Lawrence, KS 66044

William Davis
Department of Mathematics
Ohio State University
Columbus, OH 43210

Sandra Dawson
851 Echo Lane
Glenview, IL 60025

Frank Demana
Department of Mathematics
Ohio State University
231 West 18th Avenue
Columbus, OH 43210

John Dossey
RR 1, Box 165
Eureka, IL 61530

Edward Dubinsky
Department of Mathematics
Purdue University
West Lafayette, IN 47907

Wade Ellis
4652 Alex Drive
San Jose, CA 39510

Susanna Epp
Department of Mathematics
DePaul University
2219 N. Kenmore
Chicago, IL 60614

James Fey
Department of Mathematics
University of Maryland
13804 Beacon Hollow
Wheaton, MD 20906-3002

James Fife
Department of Mathematics and Computer Science
Lincoln University
Lincoln University, PA 19352

Martin Flashman
Department of Mathematics
Humboldt State University
Arcata, CA 95521

Paul Foerster
7 Sissinghurst
San Antonio, TX 78209

Susan Forman
Math Science Education Board
National Research Council
2101 Constitution Ave., NW
Washington, DC 20418

Sol Garfunkel, Director
COMAP Incorporated
57 Bedford Street #210
Lexington, MA 02173

Andrew Gleason
110 Larchwood Drive
Cambridge, MA 02138

John Goebel
Department of Mathematics
North Carolina School of Math and Science
1219 Broad Street
Durham, NC 27701

Sheldon Gordon
Department of Mathematics
Suffolk County Community College
Selden, NY 11784

Richard Griego
Department of Mathematics
Northern Arizona University
Flagstaff, AZ 86001-5717

Deborah Hughes Hallet
Department of Mathematics
Harvard University
Cambridge, MA 02138

Charles Hartford
Mathematics Editor
DC Heath & Company
125 Spring Street
Lexington, MA 02173

John Harvey
Department of Mathematics
University of Wisconsin
480 Lincoln Drive
Madison, WI 53706-1388

Kathleen Heid
Department of Mathematics Education
171 Chambers Building
Pennsylvania State University
University Park, PA 16802

JoEllen Hillyer
153 Cypress
Brookline, MA 02146

Chancey Jones
Educational Testing Service
25 E. Test Development Division
Princeton, NJ 08541

John Kenelly
327 Woodland Way
Clemson, SC 29631

Douglas Kurtz
Department of Mathematical Sciences
New Mexico State University
Las Cruces, NM 88003

Donald LaTorre
Department of Mathematics
Clemson University
Clemson, SC 29631

Katherine Layton
Department of Mathematics
Beverly Hills High School
16566 Chattanooga Pl.
Pacific Palisades, CA 90272

D. J. Lewis
Department of Mathematics
University of Michigan
Ann Arbor, MI 48109

David Lomen
Department of Mathematics
University of Arizona
Tucson, AZ 85715

Christine Lucas
Department of Mathematics
Whitefish Bay High School
1200 E. Fairmount
Whitefish Bay, WI 53217

JoAnn Lutz
N. C. School of Science and Math
Box 2418
Durham, NC 27715

Carolyn Mahoney
Department of Mathematics
California State University, San Marcos
820 W. Los Vallecitos Blvd.
San Marcos, CA 92069

Jerold Mathews
Department of Mathematics
Iowa State University
Ames, IA 50011

John McConnell
Department of Mathematics
Glenbrook South High School
4000 West Lake Avenue
Glenview, IL 60025

Paul McCreary
Department of Mathematics
University of Illinois
1409 West Green Street
Urbana, IL 61801

William Medigovich
Department of Mathematics
Redwood High School
1 El Sereno Court
San Francisco, CA 94127

Robert Megginson
Department of Mathematics
University of Michigan
317 West Engineering
Ann Arbor, MI 48109-1092

Donal O'Shea
Department of Mathematics
Mt. Holyoke College
Clapp Lab
South Hadley, MA 01075

Arnold Ostebee
Department of Mathematics
St. Olaf College
Northfield, MN 55057

Charles Patton
Hewlett-Packard
1000 NE Circle Blvd.
Corvallis, OR 97330

Tony Peressini
Department of Mathematics
University of Illinois
1409 West Green Street
Urbana, IL 61801

Horacio Porta
Department of Mathematics
University of Illinois
1409 West Green Street
Urbana, IL 61801

Gerald Porter
Department of Mathematics
University of Pennsylvania
209 South 33rd Street
Philadelphia, PA 19104

Janet Ray
Department of Mathematics
Seattle Central Community College
P.O. Box 452
Preston, WA 98050

Peter Renz
101 Colchester Street
Brookline, MA 02146

Diane Resek
8 Poppy Lane
Berkeley, CA 94708

Stephen Rodi
2008 Lazybrook
Austin, TX 78723

William Rudolph
Department of Mathematics
Iowa State University
Ames, IA 50011

James Sandefur
7301 Tyler Avenue
Falls Church, VA 22042

Doris Schattschneider
Department of Mathematics
Moravian College
1200 Main Street
Bethlehem, PA 18018-6650

Christopher Schaufele
Department of Mathematics
Kennesaw State College
PO Box 444
Marietta, GA 30061

Ray Schiflett
2013 East Union Avenue
Fullerton, CA 92631

Donald Small
Department of Mathematics Sciences
US Military Academy
West Point, NY 10996

David Smith
Co-Director, Project CALC
Department of Mathematics
Duke University
Durham, NC 27706

Horace Smith
Department of Mathematics
Southern University
7768 Emile Street
Baton Rouge, LA 70813

Richard Stanley
Dana Center
University of California
230 B Steven Hall
Berkeley, CA 94720

Karel Stroethoff
Department of Mathematical Science
University of Montana
Missoula, MT 59812

Keith Stroyan
Department of Mathematics
University of Iowa
Iowa City, IA 52242

Elizabeth Teles
Division of Undergraduate Education
National Science Foundation
4201 Wilson Blvd.
Arlington, VA 22230

ELias Toubassi
Department of Mathematics
University of Arizona
Tuscon, AZ 85721

Shirley Treadway
1609 West Walnut
Robinson, IL 62454

Alan Tucker
Applied Mathematics Department
SUNY at Stony Brook
Stony Brook, NY 11794

Thomas Tucker
Department of Mathematics
Colgate University
Hamilton, NY 13346

J. Jerry Uhl
Department of Mathematics
University of Illinois
1409 West Green Street
Urbana, IL 61801

Zalman Usiskin
Director, UCSMP
University of Chicago
5835 South Kimbark
Chicago, IL 60637

Bert Waits
Department of Mathematics
Ohio State University
231 West 18th Avenue
Columbus, OH 43210

Franklin Wattenberg
Department of Mathematics
University of Massachusetts
Amherst, MA 01003

Lee Yunker
Department of Mathematics
West Chicago High School
233 North Oak Street
West Chicago, IL 60185